THE ANIMALS OF ANCIENT EGYPT

THEN AND NOW

Richard Hoath and Salima Ikram

The American University in Cairo Press
Cairo New York

DEDICATIONS
RH: To Liz
SI: To Brian, Richard, Nicholas, Sir David Attenborough, and all creatures, great & small

This edition first published in 2026 by
The American University in Cairo Press
113 Sharia Kasr el Aini, Cairo, Egypt
420 Lexington Avenue, Suite 1644, New York, NY 10170
www.aucpress.com

Photo Credits
Front cover: Hunting scene from the Tomb of Idut, Saqqara © S. Ikram; bottom left Hippopotamus © CC H. Cornelissen; bottom right Hippopotamus from the Tomb of Idut, Saqqara © S. Ikram
Title page: Kites mourning the god Osiris from the Tomb of Nefertari, Luxor © K. Amaury
Contents page: Egyptian Goose © CC Dooncoombez

ISBN 978 1 649 03350 5

Library of Congress Cataloging-in-Publication Data forthcoming

1 2 3 4 5   29 28 27 26 25

Edited by Krystyna Mayer
Designed by Nigel Partridge

Printed and bound in Malaysia by Times Offset (M) Sdn. Bhd.

·Contents·

Introduction 4

Biogeography of Egypt 7

Human Development 13

Ancient Egypt & Its Wildlife 14

Modern Developments 16

Egypt's Bird Migrations 19

Chronology of Ancient Egypt 21

Gods & Goddesses of Ancient Egypt 22

Glossary 24

Using this Book 30

Modern & Ancient Species Descriptions 32

References & Further Reading 156

Index 158

Acknowledgements 160

Introduction

Of the great ancient civilizations, that of pharaonic Egypt is without doubt the longest lasting, best preserved and most famous. From Menes uniting Upper and Lower Egypt in around 3000 BC to the death of Cleopatra in 30 BC the civilization lasted for three millennia. But long before Menes, the inhabitants of what is now Egypt had left their mark. Deep in the Western Desert, in the Gilf Kebir on the modern state's southwestern border with Libya, rock art from 8,000 years ago attests to people living in an environment very different from that of today, a savannah environment. This is echoed throughout the Western Desert. On the walls of cliffs and caves are images of animals now long gone or almost gone, of elephants, giraffes, oryx and other species of antelope, as well as ostriches. These are creatures you now would have to go to the plains of East Africa to find. In addition to the wild animals, the domesticates, particularly cattle, were shown in rock art.

The climate changed and heated, and the lakes and rivers dried up, the grasslands retreated and human and beast alike gravitated towards the only major source of fresh water, the Nile and its Valley. After periods of pre-Dynastic experience and experiment, the ancient Egyptian civilization evolved from the historical First Dynasty onwards through to the Ptolemies and that last Queen. Today, the relics of those millennia are attested by the tombs and temples, pyramids and archeological sites unrivalled to this day. Visitors today flock to view those sites from Alexandria south, through Giza and Saqqara, to Dahshur and Meidum, and on up the Nile. At Luxor, from the vast temple of Karnak to the iconic Valley of the Kings, the pharaohs

S. Ikram

In the background is the Step Pyramid of Djoser, the first pyramid to be built and in the foreground, the Valley Temple of Unas, at Saqqara.

A. -M. Kellen

A small dog that was mummified and probably given as a votive offering to Anubis

left their mark. Further south too, to Esna, Edfu, Kom Ombo and on to Aswan and up to Abu Simbel, the civilization left its indelible footprint. While these monuments in what is now modern Egypt provide testimony enough to this incredible civilization, in museums and collections in Egypt and worldwide are the mummies, papyri, statues and statuettes, and other artefacts that are all part of this incredible legacy.

All of these are in themselves of intense fascination to the Egyptologist, historian and now the vast numbers of tourists attracted by this legacy. But for the naturalist there is a special allure. The ancient Egyptians did not just build and hew and paint and mummify and adorn, but were close observers of nature. Indeed, animals were key not only to their physical survival, providing food, but to their spiritual and cultural existence, as their gods and language were all inspired by nature. The Egyptians recorded with often astonishing accuracy the animals that existed in the world around them. Today, the modern visitor can visit Egypt's ancient sites and find on the tombs and temple walls representations of these animals depicted with such accuracy that the naturalist can often identify to species level the animals that are portrayed.

In this book, on the left-hand side of each species entry is a description of the living animal and its habits and habitat, and its current status in Egypt and elsewhere, accompanied by a photograph of it and, where appropriate, of similar species. On the right-hand side is a description of the animal in ancient Egypt, with its status and importance and depictions of it from the tombs and temples. So, the modern visitor can go to the temple at Kom Ombo north of Aswan and see the depictions of the god Horus, the falcon-headed

S. Ikram

At Kom Ombo temple the gods Horus (raptor) and Sobek (crocodile) were worshipped

S. Ikram

Cattle and their human caretakers attest to the importance of animals to humans from the dawn of Egyptian civilization

god, and at the same time, perhaps see the living species, the Common Kestrel that nests on the same temple today. Kom Ombo is unique among ancient Egyptian temples in that it was dedicated to two deities, the other being Sobek, the crocodile-headed god of the Nile. To see the Nile Crocodile today, the visitor would have to travel further south where the living entity is still present in the artificially created Lake Nasser behind the Aswan High Dam. At Kom Ombo, too, there is a small museum describing the importance of Sobek, with superb exhibits of huge crocodile mummies and other representations. The temple also boasts hieroglyphs of Barn Owls, there at night, and Northern Lapwings, still present now as winter visitors.

At Saqqara, the Old Kingdom tombs of Mereruka, Idut and Ti have images of long-gone papyrus swamps with hippos and crocodiles. On the stylized papyrus stems and umbels are Purple Swamphens, Sacred Ibises, Little Egrets, Egyptian Mongooses, possibly genets and a solitary depiction of an otter. Such is the accuracy of the portrayals that the modern naturalist can identify the Pied Kingfisher not just as a realistic representation of the bird's appearance but also of its habits, since it is portrayed in its distinctive hovering hunting mode. The sexes can even be distinguished, the male with two breast-bands and the female with one. After seeing these representations from nearly 5,000 years ago, the visitor can go back to Cairo and see the living birds looking and behaving just the same on the Nile in twenty-first century central Cairo.

This experience is repeated throughout Egypt – the monuments of the Nile Valley are well known, but also in Sinai, the Eastern and Western Deserts, including the remote oases as far west as Siwa, and south to the Gilf Kebir and even to Gebel Uweinat, where the rocks are covered with petroglyphs of giraffes now gone from modern Egypt.

This book is therefore a celebration of the incredible role of animals in ancient Egypt – as deities, as animals of domestic or wild importance, and even as pets. But it is also a celebration of the animals of modern Egypt. True, some have disappeared due to natural factors or through habitat alteration or persecution. However, many remain and can be seen to this day and identified from the amazing and often stunningly accurate representations from many thousands of years ago. Arguably, this type of book would not be possible with any other ancient civilization.

BIOGEOGRAPHY OF EGYPT

The modern state of Egypt is positioned in the northeastern corner of the African continent. Egypt today has an area of c. 1,019,449km^2. The significance of this area is that in biogeographical terms it is not part of the Western Palearctic, so some Afro-tropical species make their way into Egypt. The following numbers of species have been recorded from this modern entity: some 466 birds, 100 mammals, and 70 reptile and amphibian species, as well as nearly 50 freshwater fish and around 1,000 marine fish in the Red Sea alone, 61 butterflies and innumerable other invertebrate species. Across all groups there are few endemics, largely due to the mainly artificial borders of the current state, based on political agreement rather than natural boundaries such as rivers or mountains.

Today, Egypt is a predominantly desert country, the frequently quoted figure being between 95 and 96 per cent desert. The country is split into two. To the west is the Western Desert and to the east the Eastern Desert extending to Sinai, a triangular peninsula of some 61,000km^2 to the east of the Red Sea, rather undulating and uniform in the north with spectacular granitic mountains in the south. The Eastern and Western Deserts are bisected by the Nile River, the only watercourse of any size or consequence in the country and, save for a few isolated oases, its only source of

S. Ikram

The Nile is Egypt's life blood, providing sustenance and transport fresh water. The Nile flows north from the Sudanese border through Lake Nasser, whose waters are held back by the Aswan High Dam. From Aswan it continues north in the valley that bears its name to just north of Cairo. Here it splits into two branches (five in antiquity), the Rashid (Rosetta) branch to the west and the Damietta branch to the east. These branches diverge, and the resulting triangle of the Nile Delta is now Egypt's bread basket and home to the vast majority of its human inhabitants and including its second city, Alexandria.

Separating the Eastern and Western Deserts is the Nile Valley, referred to here simply as the Valley, which runs its way north via a large curve at modern Qena to the great fan of the Nile Delta, referred to here as the Delta. This is a landscape almost entirely moulded by humans, having been settled for much of the previous 7,000 years. Very few areas of natural Nilotic vegetation remain, most notably on Ghazal and Saluga Islands just to the south of Aswan, and to the north of the High Dam. South of the High Dam, the Valley has been flooded by the human-made creation of Lake Nasser. Today, this lake supports not only Egypt's last population of Nile Crocodiles *Crocodylus niloticus*, but also large populations of wintering waterfowl, waders and the bulk of Egypt's breeding population of the Egyptian Goose *Alopochen aegyptiacus*. In recent decades, the African Pied Wagtail *Motacilla aguimp* has established itself as a breeding resident on the lake shores, with other African species such as the Pink-backed Pelican *Pelecanus rufescens*, Yellow-billed Stork *Mycteria ibis* and Namaqua Dove *Oena capensis*. The African Golden Wolf *Canis lupaster* – once known as the Common or Golden Jackal – has been recorded from the shores of the lake, while the Nile Perch *Lates niloticus*, weighing up to 70kg, supports a local sport fishing trade. The Nile Monitor *Varanus niloticus* is a large lizard species, up to 2m long, which is regularly recorded on the shores of Lake Nasser, with a range that extends north at least to Luxor, if not on to Sohag.

North of Aswan, the Valley and Delta support an avifauna in some ways familiar to a visitor with European experience, but in other ways distinctive. The House Sparrow *Passer domesticus* is one of the most common species in and around Cairo, but then so is the Common Bulbul *Pycnonotus barbatus*, and while the Eurasian Collared Dove *Streptopelia decaocto* has established a foothold in Egypt, the common urban dove is the Laughing Dove *S. senegalensis*.

Nile Monitor

A number of African species make their way down the

Nile to Egypt. One of the characteristic birds of the river is the Pied Kingfisher *Ceryle rudis*, readily seen hovering just as depicted by the ancient Egyptians. Another kingfisher, the White-throated Kingfisher *Halcyon smyrnensis*, is a recent colonist that has rapidly expanded its range west to Fayoum and south at least as far as Minya, with unconfirmed reports from Luxor. Other African species include the Senegal Thick-knee *Burhinus senegalensis*, Kittlitz's Plover *Charadrius pecuarius* and Senegal Coucal *Centropus senegalensis*. In the countryside, the Black-winged Kite *Elanus caeruleus*, Little Green Bee-eater *Apiaster orientalis* and Crested Lark *Galerida cristata* are familiar species, while the Delta is a stronghold for the secretive Painted Snipe *Rostratula benghalensis*.

B. Grewal

Senegal Coucal

Egypt has two coastlines. The northern Mediterranean coastline stretches from Salum in the west on the Libyan border, east to Alexandria and across the coastal lakes of the northern Delta, and further across north Sinai. The coastal strip, between Salum and Alexandria, is of particular importance. It is cooler and has higher rainfall than the rest of the country, and supports a distinctive flora and fauna. Some bird species, such as the Barbary Partridge *Alectoris barbara* and Dupont's Lark *Chersophilus duponti* may be locally extinct due to hunting and habitat destruction, but others remain, such as the Common Raven *Corvus corax*, the rare Houbara Bustard *Chlamydotis undulata* and the very local Thekla Lark *Galerida theklae*. This area of coastal desert may still support populations of the very rare Egyptian Tortoise *Testudo kleinmanni*, as well as the endangered Four-toed Jerboa *Allactaga tetradactyla*. This jerboa is found elsewhere only across the border in eastern Libya. Another more widespread species is the Long-eared Hedgehog *Hemiechinus auritus*. Like all the species along the North Coast, it is threatened by habitat destruction caused by extensive tourist development over recent years.

Further east, the coastline between the northern Delta and on across northern Sinai is dominated by a series of saline lakes that support large numbers of wintering and migratory birds. These lakes, such as Maryut, Idku, Burullus and Manzala, and on further east to Bardawil in northern Sinai, support a rich avifauna hosting large

St Catherine's Monastery viewed from Mount Sinai

flocks of waterfowl and waders in winter, along with gulls including Audouin's Gull *Larus audouinii* and tern species, egrets and herons, and smaller numbers of Greater Flamingos *Phoenicopterus roseus* (also found in Fayoum). On migration, and especially in autumn, these same areas, as first landfall for birds crossing the Mediterranean heading south, offer fabulous birdwatching opportunities. Sadly, there is a long history of bird netting across the North Coast, ostensibly for Common Quail *Coturnix coturnix*, but the nets do not discriminate and each year millions of birds die on passage.

Egypt is contiguous with the Middle East, or was until the completion of the Suez Canal in 1869 through the Sinai Peninsula, an inverted triangle of land some 61,000km^2 in area. South of the coastal lakes mentioned above, the north is relatively low lying, with sandy plains and low hills, while a dominant feature of the south is the mountainous area around St Catherine, famous for its monastery and Mount Sinai (although at 2,641m, Gebel Katherina is the highest point). Characteristic birds of these mountains include the Chukar *Alectoris chukar*, White-spectacled Bulbul *Pycnonotus xanthopygos*, Palestine Sunbird *Cinnyris osea*, Tristram's Starling *Onychognathus tristramii* and the Sinai Rosefinch *Carpodacus synoicus*. In the north-easternmost corner of Sinai, a number of species, such as the Syrian Woodpecker *Dendrocopos syriacus*, Great Tit *Parus major* and Serin *Serinus serinus*, more widespread elsewhere in the region, just reach into Egypt in and around Rafah. Mammals include the elusive Sand Cat *Felis margarita*, Fennec Fox *Vulpes zerda* and, in the southern mountains, the Striped Hyena *Hyaena hyaena*. Historically, these highlands were home to the now extinct Sinai Leopard *Panthera pardus jarvisi*, although the Bedouin leopard traps that can still be seen are testament to its former presence. Distinctive reptiles include Inne's Cobra *Walterinnesia aegyptia* and the Sinai Banded Snake *Coluber sinai*, while to the north the tree frog *Hyla savignyi* creeps into Egypt. Sinai has two endemic species of butterfly, *Pseudophilotes sinaicus* and *Satyrium jebelia*. The former, known as the Sinai Baton Blue and vying for the title of the world's smallest butterfly, is now threatened with extinction due to habitat loss.

The other coastline borders the Red Sea, which is very different from that of the Mediterranean. It runs some 800km from the Sudanese border north to Hurghada, where it splits into the western Gulf of Suez and the eastern Gulf of Aqaba. In terms of biodiversity, the Red Sea supports some of the richest coral reefs in the world, as well as extensive mangroves. Bird species include breeding populations of the near endemic White-eyed Gull *Larus leucophthalmus* and Sooty Gull *L. hemprichii*, terns such as the Caspian Tern *Hydroprogne caspia*, White-cheeked Tern *Sterna repressa*, Lesser Crested Tern *S. bengalensis*, Crested Tern *S. bergii* and Bridled Tern *Onychoprion anaethetus*, the Brown Booby *Sula leucogaster* and one of the densest breeding populations of the Osprey *Pandion haliaetus*. At sea, cetacean species such as the Bottle-nosed Dolphin *Tursiops truncatus*, Spinner Dolphin *Stenella longirostris*, Pan-tropical Spotted Dolphin *S. attenuate* and Risso's Dolphin *Grampus griseus* can be found, as well as the endangered sirenian, the Dugong *Dugong dugon*.

However, modern Egypt is still predominantly desert, and the popular perception of the desert is of a barren, lifeless environment – desert is desert, is desert. There are elements of truth in this. Parts of Egypt's Western Desert are among the hottest and driest places on the planet, and life is indeed very difficult to find. However, the desert is an extraordinary and extraordinarily diverse and fragile environment

S. Ikram

The deserts bracket the Nile Valley and are home to a variety of fauna inhabited by plants and animals that have evolved to survive its harshness and challenges. It is a varied landscape. The Western Desert is a vast expanse of relatively low-lying, flat hamada desert – indeed the Qattara Depression to the north lies up to 134m below sea level. Running diagonally across the Western Desert from north-west to south-east is an extensive area of duneland known as the Great Sand Sea, which is hot and arid, with some of the driest and hottest places on the planet, but even here life persists. Characteristic species include the Desert and Bar-tailed Larks *Ammomanes deserti* and *A. cincture* and the White-crowned Wheatear *Oenanthe leucopyga*, while mammals include both the Dorcas and Slender-horned Gazelles *Gazella dorcas* and *G. leptoceros*, a myriad of gerbil, jird and jerboa species and, sadly extinct within the past century and a half, the Scimitar-horned Oryx *Oryx dammah*, Addax *Addax nasomaculatus*, Hartebeest *Alcelaphus buselaphus* and Cheetah *Acinonyx jubatus*. There are reptiles, too, perhaps the most remarkable being the Saharan Sand Snake *Psammophis aegyptius*, which lives in the most austere and hostile regions of the desert. Here it relies exclusively on the migration of birds in spring and autumn for sustenance.

Scattered across the Western Desert is a series of oases, Kharga, Dakhla, Farafra, Bahariya, and to the west, Siwa, referred to in this book as the Western Oases. These outposts, plus many much smaller and isolated oases, including acacia groves in the Qattara Depression, support populations of breeding species, and during spring and autumn attract legions of migrating birds. At these times of year any pocket of vegetation in the vastness of the desert is worth checking out for migrants. Gazelle are still to be found in Kharga, as are foxes (Fennec Fox and Red Fox *Vulpes vulpes aegyptiaca*), jerboas and a variety of snakes.

To the east, the Eastern Desert is very different. This is mountainous desert that runs north from the border with Sudan to Suez and is riddled east to west by deep wadis, of which Wadi Degla just outside Cairo is one of the most northerly. While some species are shared with its western counterpart, distinctive birds of the Eastern Desert include breeding Egyptian Vultures *Neophron percnopterus*, Lappet-faced Vultures *Torgos tracheliotus*, Scrub Warblers *Scotocerca inquieta* and Trumpeter Finches *Bucanetes githagineus* (also found in the Western Oases). Mammals include the Golden Spiny Mouse *Acomys russatus*, Nubian Ibex *Capra ibex* and Rock Hyrax *Lepus capensis*. In the very southeasternmost corner of Egypt's Eastern Desert, a number of species found nowhere else in Egypt breed on and around Gebel Elba, including the Shining Sunbird *Nectarina habessinicus* and Fulvous Babbler *Turdoides fulvus*. Mammals that just scrape onto Egypt's list here include the striking black and white mustelid, the Zorilla *Ictonyx striatus*, and the enigmatic Aardwolf *Proteles cristatus*. New discoveries are still being made, and in January 2025 a Spotted Hyena *Crocuta crocuta* was filmed and photographed in Gebel Elba far out of its usual habitat.

Human Development

While this book focuses on the fauna of Egypt during the late Predynastic and Dynastic Periods stretching from around 3300 BC to the death of Cleopatra in 30 BC, evidence of human activity significantly predates this. In the remoteness of the Western Desert, Stone Age tools and fragments of ostrich eggs are strewn across the desert, and to the south-west around the massif of the Gilf Kebir south to Gebel Uweinat there are extensive rock carvings and paintings dating back to some 8,000–10,000 years ago. The best known of these is at the Cave of the Swimmers, made famous in the novel and film *The English Patient*. Here, a series of humans is depicted in attitudes that have been interpreted as swimming; at the time when the paintings were executed this area was primarily savannah with grassland and lakes. Of more interest to the naturalist is the nearby Foggini-Mestikawi Cave, also known as the Cave of the Beasts. The rock face is covered with thousands of paintings of people and the animals they lived with and hunted such as giraffes, gazelles, antelopes, ostriches and curious headless creatures that have been variously identified as lions, baboons and even an early form of the sky goddess Nut. Further south at Uweinat there are more depictions, including giraffes, ostriches and a cheetah, but also domestic animals including cattle and curiously even camels. Curious, because in historical times the first evidence of camels did not appear until around 740 BC, or possibly later.

Today, this region of the Western Desert is one of the hottest and driest places on the planet. Around 6,000–7,000 years ago the climate warmed and dried, and civilization became concentrated along the Nile Valley and in the Delta and oases, with the Dynastic Period beginning with the unification of Upper and Lower Egypt by King Menes in around 3100 BC.

The Nile Valley and Delta have perhaps been continually populated by people longer than anywhere else, and their flora and fauna today are very different from that first found by the ancients. As a result of thousands of years of human impact and action, little remains of the original biome. On the river itself, at Aswan, one last area of the

S. Ikram

Humans in different attitudes in the Foggini-Mestikawi cave.

original Nilotic ecosystem can be found at the Saluga and Ghazal Protected Area. Further south above the Aswan High Dam, the Valley has been swallowed by the creation of Lake Nasser, while in the north 5,000 years of continuous human activity have created a largely manufactured environment dominated largely by non-native plant and animal species. Most of the major agricultural crops are alien – potatoes, tomatoes and maize from the Americas, rice originally domesticated in China, sugar cane from Southeast Asia, and mangoes and bananas from South and Southeast Asia. While Nilotic *Acacia* species may still be seen on Saluga and Ghazal Islands, with five species recorded, the most frequent trees now found by the canals and waterways of the Valley and Delta are *Eucalyptus* species native to distant Australasia. The fields are grazed by the domestic descendants of wild sheep, goats, cattle and donkeys but, as stated above, the camel is a relatively recent addition to the pastoral scenes, as is the Water Buffalo (approximately in the twelfth century AD or later). Dogs have been here for millennia, but the Domestic Cat, popularly thought to have been domesticated in ancient Egypt, may also have been first domesticated elsewhere.

Ancient Egypt & Its Wildlife

Fortunately, the ancient Egyptians' tombs contained images not only of their deities, but also of their own daily activities and the landscape that they inhabited. The Old Kingdom tombs at Saqqara, such as those of Idut and Ti, depict hunting and fishing scenes from those times. Some of the wildlife in these images is familiar to a modern audience, such as the herons, egrets, wildfowl and gallinules, often portrayed so accurately that they can be identified to species level by a modern visitor. Pied Kingfishers hover over the papyrus and lotus wetlands just as they do today, so clearly depicted that male and female can be differentiated, the former with one breast-

S. Ikram

A marsh scene from the tomb of Mereruka at Saqqara where it is posssible that mongooses were trained to raid birds' nests and to capture these creatures

S. Ikram

Fishing with nets or with hooks was a popular sport as well as a way of acquiring food. Several different types of fish can be identified in such scenes

band, the latter with two. That said, identification is not always so straightforward, such as with the controversy over the Meidum Geese from just a few kilometres further south. But other species have disappeared, most notably the Sacred Ibis *Threskiornis aethiopicus*. Once worshipped as Thoth, god of wisdom, writing and scribes, the last record of this species in modern Egypt is from 1891. Habitat destruction and persecution probably did for it. With the changing ecology of Lake Nasser and the creation of new habitats, there may be hope of a return. There has even been talk of actively reintroducing the Sacred Ibis into modern Egypt, although this would be controversial as it could upset the current ecosystem.

A. -M. Kellen

An elaborately wrapped crocodile mummy

Ironically, the other animal associated with Thoth and also bearing the moniker 'sacred', the Sacred or Hamadryas Baboon *Papio hamadryas*, is also extinct in modern Egypt, and probably never flourished here. Like the ibis and crocodile, it too was mummified by the ancient Egyptians and particularly well-preserved mummies were buried in Saqqara and Tuna el-Gebel near the modern city of Minya.

Prominent among the mammals in these Old Kingdom tombs is the Hippopotamus *Hippopotamus amphibius*. This too has disappeared, with the last accepted record being from 1819. It was hunted for its meat and ivory and to protect crops and people, although its ultimate demise was probably due to the introduction of modern firearms and game hunting. One depiction in the Tomb of Idut shows a scene with a noble hunting in a papyrus swamp full of hippos. One of the hippos is giving birth, with the head of the young hippo emerging from the mother's nether regions. In a rather macabre twist, there is a large Nile Crocodile waiting behind the hippo for a very fresh meal – if not the calf, the afterbirth (p. 141).

The Nile Crocodile was found throughout the Nile Valley and Delta. It was worshipped as the god Sobek and was associated with the Nile and its annual inundations, the sun and fertility. Not only was it portrayed in hunting scenes, but also it was mummified as the incarnation of or offerings to the god Sobek. Today at the Temple of Kom Ombo, a temple uniquely dedicated to two gods, Horus (see falcon, p.115) and Sobek, there is a museum containing many crocodile mummies. The Egyptian Museum's animal mummy room/room of the natural world, contains a huge, mummified crocodile which, when examined, was found to have baby crocodiles in its mouth. Modern research on living crocodiles has shown that the mother crocodile not only guards the nest of eggs from predators such as monitor

R. N. Hewison

Lake Qarun

lizards and mongooses, but that on hatching she carries the young crocodiles down to the water in her mouth. Was this crocodile taking her youngsters into the afterlife? The Nile Crocodile is still found in modern Egypt, but today is confined to Lake Nasser, where there is a population of several thousand thriving on the lake's population of Nile Perch. It has long since disappeared below the Aswan High Dam, although there are still occasional reports. These most likely refer to the Nile Monitor *Varanus niloticus*, which can approach two metres in length.

One of the centres of worship of the Nile Crocodile was the 'oasis' of Fayoum. Once an offshoot of the Nile with its waters flowing from and back into the river, today Fayoum is a dead end with the water flowing into Lake Qarun and the excess going into the two artificially created lakes of Wadi Rayan, connected by one of Egypt's very few waterfalls. The crocodile was worshipped in Fayoum from the Old Kingdom through to Roman times. It was reared in captivity at a number of sites in Fayoum, such as Karanis, Qasr Qarun, Crocodilopolis and most notably Medinat Madi. Crocodiles have long since disappeared from Fayoum, and with no natural outlet any more, Lake Qarun has changed from a freshwater to a saltwater lake. Today, it is an important wintering ground for a number of wildfowl and wader species. The Greater Flamingo *Phoenicopterus roseus* is a winter visitor, often in hundreds, and there is evidence that it is now breeding there. The Demoiselle Crane *Anthropoides virgo* is a very rare visitor to Egypt, and has also been recorded there recently, something of a homecoming as it is beautifully portrayed along with Common Cranes *Grus grus* in the tomb of Ti at Saqqara.

MODERN DEVELOPMENTS

The construction of the High Dam and the formation of Lake Nasser have created a number of artificially flooded wadis known as *khors*, some of which are permanently under water. Aside from creating suitable habitat for the Nile Crocodiles, a number of bird species have colonized the lake from the south. Notable among these are the Pink-backed Pelican *Pelecanus rufescens*, Yellow-billed Stork *Mycteria ibis*, African Skimmer *Rhynchops flavirostris* and African Pied Wagtail *Motacilla aguimp*. This new habitat may also tempt back the Sacred Ibis mentioned above, and another bird named the Egyptian Plover *Pluvianus aegyptius*. This striking wader could perhaps be named the Non-Egyptian Non-Plover as it was last recorded in Egypt in 1932 and is more closely related to the pratincoles and coursers than to the plovers. The Egyptian Plover, boldly marked in dove-grey, black and orange-infused cream, was recorded by Herodotus, who referred to it as the 'running bird'. Its other name, Crocodile Bird, refers to its supposed habit, noted by Herodotus, of picking leeches and other parasites from the open jaws of crocodiles.

The vast papyrus swamps that once covered much of the Nile Delta and Valley had largely disappeared by the ninth century AD as the swamps were drained and cleared for agriculture and settlement, and for mulberry and other paper sources. While isolated pockets of papyrus were noted through the nineteenth century, it is uncertain whether wild papyrus survived into the twentieth century. However, in the 1960s, Dr Hassan Ragab reintroduced papyrus in the vicinity of Cairo for botanical, recreational (touristic) and economic reasons. Today, Dr Ragab's Pharaonic Village in Giza provides an opportunity to get an idea of what the original papyrus swamps must have been like. Some of the denizens of those papyrus swamps have survived through to modern Egypt by adapting to the changing biosphere. One such species is the Egyptian Mongoose, or Ichneumon, *Herpestes ichneumon*. This was portrayed in the friezes clambering around in the papyrus umbels in search of birds' eggs and nests. Today, the Egyptian Mongoose can still be found over much of the Delta and northern Valley and also Fayoum, where it has recently been filmed. Another slender carnivore arguably portrayed in the papyrus is the genet. The Small-spotted Genet *Genetta genetta* has only been recorded in Egypt from Gebel Elba and the southeastern shore of Lake Nasser.

Another inhabitant of these swamps was the Swamp Cat *Felis chaus*. This is larger and lankier than the Domestic Cat *F. catus*, with a proportionately shorter tail and small ear-tufts. A predominantly Asiatic species, its only Africa foothold is in Egypt, where it extends through the Delta south at least to Luxor and also into Fayoum where, like the Egyptian Mongoose, it has recently been filmed. The Swamp Cat has successfully substituted its

Shutterstock/A Nwera

The High Dam at Aswan

natural papyrus habitat for modern maize and sugar-cane plantations. At one point the Swamp Cat was thought to have been involved with the domestication of the cat, although this is now widely refuted. However, its mummified remains have been found in limited numbers among cat mummies.

The African Wild Cat *F. lybica* is still found in modern Egypt, where hybridization with the Domestic Cat is thought to be its main threat. The cat, probably the Domestic Cat, was worshipped in ancient Egypt as the goddess Bastet. One of the main centres for the worship of Bastet was Tel Basta in modern-day Zagazig in the Delta. The cat was long thought to have been domesticated by the ancient Egyptians, although this is debated. One of the difficulties is that the tabby-type of the Domestic Cat is superficially very similar to the African Wild Cat. There are many friezes depicting cats from ancient Egypt, such as the stunning paintings from the tomb-chapel of Nebamun from 1350 BC. The difficulty is that the cat is portrayed in a wild environment, in this case a hunting scene with other wild animals and birds, and it is difficult to know whether this is a Domestic Cat, an African Wild Cat, or indeed a Swamp Cat. It is only when the cats are portrayed in a domestic context such as sitting below a chair or throne, or wearing a collar, that there is evidence, if not of domestication, then at least of taming.

If the Swamp Cat and the African Wild Cat have survived into the twenty-first century, other cat species portrayed or even revered by the ancient Egyptians have not fared so well. The Lion *Panthera leo* disappeared in historical times. Males were associated with the king, and females were worshipped as Sekhmet, daughter of Ra, a warrior goddess also associated with medicine. The Leopard *Panthera pardus* clung on in the mountains of South Sinai possibly into the 1950s; the stone-built Bedouin Leopard traps can still be seen there. It may survive in the south-east of Egypt in Wadi Allaqi, where tracks have been photographed, and Gebel Elba. The Cheetah *Acinonyx jubatus* held on in the Qattara Depression in the northern Western Desert until the 1980s, but is almost certainly extinct there now.

Clearly one of the underlying themes of the above has been that habitats change through human impact and interference, and it has been noted that a number of the species revered, worshipped, deified and portrayed by the ancient Egyptians have disappeared to live on in the modern country only on those tomb and temple friezes, as mummies or as statuary.

However, despite this, new discoveries are being made, not just in the remote corners of the desert but in the Nile Delta and Valley itself. In 1993 a new species of toad, the Nile Delta Toad *Amietophrynus kassasii*, was described from the Delta south of Damietta. In Cairo itself the local race of the Weasel *Mustela nivalis subpalmata* may, after extensive genetic studies, prove to be a full species, as might the Egyptian race of the Yellow Wagtail *Motacilla flava pygmaea*. Recent introductions, intentional or accidental, include the Streaked Weaver *Ploceus manyar*, Indian Silverbill *Lonchura malacca*, Red Avadavat *Amandava amandava* and Rose-ringed Parakeet *Psittacula malacca*. Indeed, the latter may not be so recent, as it is portrayed in the stunning mosaic floor of the Roman Villa of the Birds in Alexandria.

CC O. Tausch

The Nilometer on Elephantine Island

Other changes may have benefitted the wildlife. One of the consequences of the High Dam has been to 'tame' the Nile. The ancient Egyptians were very conscious of the annual flooding of the river. Indeed, on Elephantine Island, near Aswan, there is a Nilometer, where the annual inundation was meticulously recorded. Each Egyptian temple contained a Nilometer in order to record the level of the inundation and to calculate tax accordingly. The construction of the High Dam has stopped the floods, with the water held back in Lake Nasser and released through turbines. This has meant that the natural fertilizing of the Valley and Delta and the agricultural land through the silt borne down largely from the highlands of Ethiopia no longer occurs, necessitating the use of chemical fertilizers. One bird may have benefitted from this. The Pied Kingfisher hunts by hovering over the water, then plunging into it, diving for fish. Over recent decades it seems to have undergone a range expansion and is now one of the easiest birds to find whilst on a Nile cruise. Scientists think that as the water of the Nile is now clear and no longer laden with silt, the kingfishers are finding it easier to hunt.

Other Nile inhabitants have not fared so well. Possibly extinct below the High Dam, the Nile Soft-shelled Turtle *Trionyx triunguis* is now probably confined to Lake Nasser.

EGYPT'S BIRD MIGRATIONS

One feature that has been a constant since ancient times has been bird migration. Egypt sits on one of the great migration corridors of the world. Every spring and autumn, millions of birds fly through Egyptian airspace on their way to and from European breeding grounds and African wintering grounds. Large birds such as eagles *Aquila* species, Common Cranes, White Pelicans *Pelecanus onocrotalus* and White Storks *Ciconia ciconia* are channelled through specific corridors. These are the big birds that rely on thermals of rising warm air to gain height, then soar rather than using active flight. In Egypt these flyways follow the Red Sea coast and the Valley, where desert mountains channel the hot air currents upwards. Ain Suhkna on the

Red Sea Coast in spring, and Gebel Zeit and the Valley in spring and autumn can prove especially rewarding for the naturalist.

One of Egypt's largest birds of prey, the Lappet-faced Vulture is a rare visitor to the south of the country and can be found at Wadi Gimal near Marsa Alam on the southern Red Sea coast. It is worthy of mention not just because of its length of over a metre and 290cm wingspan, but also because it appears on the pharaoh's brow, symbolic of Upper Egypt and the goddess Nekhbet. It is paired with the Egyptian Cobra *Naje haje*, symbolic of Lower Egypt and the goddess Wadjet, and together symbolic of their unity. A fine example is on the golden mask of Tutankhamun (p. 137).

The smaller birds pass through too in huge numbers, but on a much broader front as they rely on active rather than soaring flight. During migration any patch of greenery may reveal birds, however remote the habitat; indeed, the desert oases may prove especially rich at these times precisely because the habitat is so isolated. Not all the birds pass through, though. Many autumn migrants stop and overwinter in Egypt and the northern lakes, the Delta and Valley, Lake Qarun in Fayoum and the coasts of the Red Sea and Mediterranean. Winter is especially rich in terms of waterfowl and waders.

The ancients recorded this too. Perhaps one of the best examples is in the tomb of Khnumhotep III at Beni Hasan near Minya. Here, resident species such as the Hoopoe *Upupa epops* may be seen in beautifully portrayed garden habitat surrounded by migratory species, all strikingly and accurately rendered. They include the Masked Shrike *Lanius nubicus*, Red-backed Shrike *L. collurio*, Common Redstart *Phoenicurus phoenicurus* and European Turtle Dove *Streptopelia turtur*. In Cairo, a city of some 23 million people, one of the spectacles of spring is the fly-past of European Bee-eaters *Merops apiaster* from mid-April through May. They may be joined by Blue-cheeked Bee-eaters M. *persicus*, some of which will carry on to Europe while others may stay to breed, for example at Saqqara. Moreover, it is not just the birds that are on the move. Kuhl's Pipistrelle *Pipistrellus kuhlii*, one of Egypt's smallest bats, is known to be migratory over parts of its range, although there is no evidence from Egypt yet. And perhaps most remarkable of all is a butterfly, the Painted Lady *Vanessa cardui*, tentatively identified from the tomb-chapel of Nebamun (and other images) mentioned above. This highly migratory species has been recorded throughout Egypt from even the remotest areas of the Western Desert, including the Gilf Kebir.

In conclusion, Egypt has a surprisingly rich fauna for a country that is popularly regarded as a desert country. The fauna of the desert is highly adapted, largely nocturnal and sparsely distributed. That of the Nile now is much reduced and many of the species worshipped, revered, mummified, drawn, portrayed and sculpted by the ancients have either disappeared or are now very rare. However, through those portrayals, the ancient Egyptians have given us a singular insight into their own natural history that has, perhaps uniquely, made a book such as this possible.

CHRONOLOGY OF ANCIENT EGYPT

Early Dynastic Period

Dynasty 1	3050–2800 BC
Dynasty 2	2800–2660 BC

Old Kingdom

Dynasty 3	2660–2600 BC
Dynasty 4	2600–2470 BC
Dynasty 5	2470–2360 BC
Dynasty 6	2360–2200 BC

First Intermediate Period

Dynasties 7/8	2200–2100 BC
Dynasties 9/10	2100–2000 BC
Dynasty 11a	2080–2010 BC

Middle Kingdom

Dynasty 11b	2010–1943 BC
Dynasty 12	1943–1760 BC
Dynasty 13	1765–1650 BC

Second Intermediate Period

Dynasty 14 (LE*)	1700–1650 BC
Dynasty 15 (LE*)	1650–1525 BC
Dynasty 16 (UE*)	1660–1585 BC
Dynasty 17 (UE*)	1585–1530 BC

*LE Lower Egypt
*UE Upper Egypt

New Kingdom

Dynasty 18	1530–1278 BC
Dynasty 19	1278–1176 BC
Dynasty 20	1176–1078 BC

Third Intermediate Period

Dynasty 21	1078–941 BC
Dynasty 22	948–743 BC
Dynasty 23	834–730 BC
Dynasty 24	728–712 BC
Dynasty 25	755–656 BC

Saite Period

Dynasty 26	672–525 BC

Late Period

Dynasty 27 (Persians)	525–404 BC
Dynasty 28	404–398 BC
Dynasty 29	398–379 BC
Dynasty 30	379–340 BC
Dynasty 31 (Persians)	340–332 BC

Hellenistic Period

Dynasty of Macedonia	332–310 BC
Dynasty of Ptolemy	310–30 BC

Roman Period 30 BC – AD 395

Gods & Goddesses of Ancient Egypt

Amun A chief god of Egypt who often was syncretized with the sun god Re. Amun means 'the hidden one', and he embodied the potential of fertility and power. His main temple was the great temple at Karnak in modern-day Luxor.

Anubis Canine-headed deity in charge of mummification as well as helping the deceased travel between this world and the next. He was also protector of travellers and some scholars think that he segued into St Christopher, patron saint of travellers. No single canine has been identified satisfactorily with him, and thus it is thought that he is an amalgam of many creatures (dog, jackal – now identified as an Egyptian Wolf – fox and hyena), creating a super-canine.

Anuket A protective goddess associated with the First Cataract, the southern border of Egypt, who was also worshipped in Nubia. Her totemic animal was the gazelle.

Apademak Chiefly a Nubian deity who was leonine and associated with warfare.

Atum A primeval creator deity whose main cult place was Heliopolis. He was associated with the sun (Re-Atum) and thus with many solar creatures as well as other creatures, such as the snake, shrew, scarab, mongoose, lion, bull, lizard, ram and an ape wielding a bow.

Bastet Goddess of motherhood, love, beauty and self-indulgence, often shown as a cat. Her main cult centre was at Bubastis.

Geb The god of the earth whose totemic animal was the goose.

Hapi A god who represented the power of the Nile and its fertility. He was often shown as a man with pendulous breasts and a large stomach. He was thought to dwell in a cavern in Aswan.

Hathor Solar goddess associated with kingship and motherhood, also associated with cattle. Her main temple is at Dendera.

Heket Goddess of magic who assisted in creating the child in the womb, often shown as a frog or a frog-headed woman due to the fact that tadpoles seem to magically metamorphose into frogs.

Horus The reigning king was a manifestation of the sky/solar deity, Horus, son of Osiris (god of the Underworld) and Isis. He is shown as a raptor.

Ipet A goddess associated with childbirth, who could be shown as a female hippopotamus, standing erect on lion feet, with human arms, and with the back and tail of a crocodile.

Isis Mother goddess and great magician, wife of Osiris (god of the Underworld), mother of Horus and sister of Nephthys and Seth. She is often shown as a kite.

Khentiamentiu One of the earliest deities of the afterworld, who subsequently became associated with Osiris. His name means 'Foremost of the Westerners' and he, like Anubis, was a canine deity, a combination of a jackal (Egyptian Wolf), fox and dog.

Khepri A sun god who was the manifestation of the morning sun and shown as a

scarab beetle or a man with a scarab beetle as a head. This association was due to the fact that dung beetles roll dung balls that can be likened to the sun.
Madfet A protective deity associated with felines and mongooses. She protected people against snake and scorpion attacks.
Mahes This leonine deity is sometimes shown fighting with the sun god against the serpent Apophis. Mahes was a warrior god and protector of sacred places.
Meretseger Deity associated with the Theban mountains and protectress of cemeteries therein. She is generally shown as a serpent.
Mut Mother goddess and consort of Amun, a chief god of Egypt. She was associated with vultures and lionesses.
Neith A deity who was associated with warfare, with a shield and arrows on her head. She was associated with creation and motherhood and was the most important deity of Lower Egypt. She was sometimes shown as a cow or a serpent.
Nekhbet Goddess of Upper Egypt, personified as a vulture. She was a royal emblem and protected the pharaoh in particular.
Nephthys Sister of Isis and Osiris (god of the Underworld), wife of Seth, a mourner sometimes shown as a kite, together with Isis.
Nut Sky goddess, associated with both the sow and the cow.
Re A chief god and a sun god associated with a variety of creatures: primarily, raptors, serpents, lions and shrews.
Renenutet Goddess of the harvest, shown as a rearing cobra. She was also regarded as the divine nurse.
Reret The name means 'The Great Sow', but the goddess is shown as a hippopotamus. She was associated with the constellation that is now known as Draco, which contains the ancient pole star, Thuban.
Reshep A west-Semitic god of thunder and war, assimilated into the Egyptian pantheon during the New Kingdom. He was associated with Nergal, the Mesopotamian god of warfare and pestilence. He is shown as a man with a crown adorned with a head or horns of a gazelle.
Ruty The twin lions associated with the Eastern and Western Deserts and the ideas of yesterday and tomorrow.
Sekhmet Fierce goddess, often depicted as a lioness or a woman with a lion's head. She was warlike and could bring as well as avert plague.
Selket or **Serket** A deity who protected against scorpions. She is generally shown as a woman with a scorpion on her head and was one of the guardians of the canopic jars that contained mummified viscera.
Seth This complex deity was the brother of Osiris, Isis and Nephthys. He commanded the desert and could control storms. His totemic animal seems to have been an amalgam of a donkey, aardvark, and possibly pig. He provided the balance to Horus, who ruled over the Nile Valley.

Sobek Crocodile-headed solar deity, sometimes shown only as a crocodile. He governed the Nile floods and was also a fertility god.
Taweret The quintessential deity associated with childbirth. Taweret was shown as a pregnant hippopotamus with pendulous breasts wearing a long wig.
Tefnut Goddess of moisture and humidity, one of the first deities to be created. She was the daughter of Atum. She was shown as a woman, a lioness or a woman with a lioness's head.
Thoth God of writing and wisdom, shown as an ibis or baboon or a man with the head of an ibis.
Tutu An apotropaic deity shown as a walking sphinx with bird wings and a serpent's tail.
Wadjet Goddess of Lower Egypt, who took the form of a cobra ready to strike. She was sometimes identified with the eye of the sun god, Re.
Wepwawet The oldest of Egypt's canine deities. He was shown as a grey canine, possibly a wolf, and antedated Anubis, who subsequently took on his attributes. His cult centre was Asyut.
Weret-Hekau 'Great Enchantress' was associated with the crown of Lower Egypt as well as a rearing cobra, sometimes with the head of a woman.

Glossary

abdomen In insects, one of the three body divisions directly behind the thorax, of up to 10 segments.
Acacia Important genus of desert/semi-desert/savannah trees.
adult Mature; capable of breeding.
aerial Making use of the open sky.
amphibians Vertebrate species of the class Amphibia, represented in Egypt only by the frogs and toads.
amulet Small object, such as a hieroglyphic symbol or figurine of a god or animal, believed to provide magical protection or other benefits to its wearer.
antenna (pl. **antennae**) Sensory feeler on the head of, for example, an insect.
aquatic Living on or in water.
arachnid Member of the class Arachnida, including spiders and scorpions.
arboreal Living in trees.
Book of the Dead Also known in ancient Egypt as the book of 'Coming Forth by Day', the Book of the Dead refers to a collection of prayers, magic formulas, and hymns to be used by the soul of the deceased for guidance and protection on its journey to the afterlife. From the New Kingdom to the Roman period, these books were written on papyrus scrolls and buried with the deceased.
callosities thickened areas of skin.

canid Dog-like mammal like a wolf, jackal or fox.
canopy Leafy foliage of treetops.
carapace Hard, shell-like structure protecting the upperparts of, for example, crabs, turtles and tortoises.
carnivore Animal that eats meat
cheek Term loosely applied to sides of head, below the eye or on ear-coverts.
cippus (pl. **cippi**) A type of protective statue, featuring the child god Horus standing on the back of a crocodile, holding lions, snakes or other dangerous animals. Water, wine or milk was poured over the cippus and then drunk, curing the drinker of any ailment.
Coffin Texts Inscriptions accompanied by vignettes painted or carved inside coffins from the Middle Kingdom onwards, whose purpose was to protect the dead and guide their transition between this world and the next.
collar Distinctive band of colour that encircles or partly encircles neck.
coverts Small feathers on wings of birds (lesser, median, greater, primary and so on), ear and tail-base.
crepuscular Active at dusk and dawn.
crest Extended feathers on head of bird or elongated scales along head and back of reptile.
crown Top of head.
crown-stripe Distinct line from forehead along centre of crown.
culmen Top edge of upper mandible.
diurnal Active by day.
dorsal fin Structure on or along back of a fish and many cetaceans.
dorsum Upperparts or back.
ear-coverts Feathers covering ear opening. Often distinctly coloured.
echolocation Particularly in bats and cetaceans, the 'sense' that locates objects or prey by reflected sound waves (sonar).
eclipse Refers to non-breeding plumage of ducks, where males often look much like females.
endemic Indigenous and confined to a particular place or region.
exoskeleton In, for example, insects, the hard outer shell of the body.
extant Still existing.
extinct No longer in existence.
eye-ring Contrasting ring around eye.
eye-stripe Stripe through eye.
faience Self-glazing, non-clay ceramic material. It was generally made from crushed quartz and used for large-scale production of jewellery and amulets, shabtis, vessels and architectural decoration, often coloured blue or green by the addition of copper.
family Specified group of genera.

felids Cat-like mammals, such as lions, leopards and African Wild Cat.
felucca Traditional small boat propelled by oars and/or lateen sails, used on the Nile, formerly more commonly.
flank Side of body.
forearm Especially in bats. Measurement between wrist and elbow.
foreneck Lower throat of a mammal.
form Broadly synonymous with subspecies.
gape Basal part of beak.
genus Group of related species.
hackles In birds, long and pointed neck feathers. In mammals, erectile hairs along back.
hamada Stony desert with rock fragments covering a sedimentary rock base; common terrain in the northern Western Desert.
hepatic Refers to rust- or liver-coloured plumage phase, mainly in female cuckoos.
herbivore Refers to animal with a diet consisting mainly of plants.
hybrid Infertile (generally) offspring of two different species.
immature Not yet adult.
insectivore Refers to animal whose diet consists largely of non-vertebrates. Taxonomically a mammal of the order Insectivora.
introduced Not native. Brought into an area by intentional or unintentional human agency.
invertebrate Broadly any animal without a backbone.
iris Coloured eye membrane surrounding pupil.
Isfet Concept of chaos and disorder, the opposite of Maat.
IUCN International Union for Conservation of Nature.
juvenile Young – of birds in first plumage, of mammals beyond infant but not fully adult.
Karnak Called by the Egyptians 'Throne of the Two Lands' and 'The Finest of Seats', it is the site of the temple of Amun at Thebes. One of the largest religious complexes ever constructed, its 250 acres are the site of temples, chapels, obelisks, statues, orchards, wells, lakes, houses and storage rooms built over the course of 2,000 years.
khor Flooded or seasonally flooded desert valley, for example on Lake Nasser.
lanceolate Lance shaped; slim and pointed.
larva In, for example, insects, juvenile stage between egg and pupa/adult (like a caterpillar).
lore(s) Between the base of bill and eye.
Lower Egypt Area north of Cairo, primarily comprising the Delta, thus called as the Nile flows from south to north.
Maat Concept of balance and rightness in life; a person had the duty to live in accordance with Maat; also, the deity by that name charged with establishing and

judging truth and honesty. The opposite is Isfet.
malar Stripe on side of throat.
mammal Member of the order Mammalia, characterized by mammary glands.
mandible(s) Each of the two parts of bill. In insects the first pair of mouthparts.
mangroves Trees of various species, such as *Avicennia* spp., forming important intertidal habitat.
mantle Back, between wings in birds.
mask Dark plumage around eyes and ear-coverts.
mastaba Tomb type, common from the Early Dynastic Period onwards. Derives from the Arabic word for mudbrick bench, which the tombs resemble.
melanistic Black colour phase, for example of the urban Eastern Spiny Mouse.
Memphite Environs of Memphis (modern Badrashin), the ancient capital of Egypt south of Cairo. Its cemeteries include Abu Rawash, Giza, Abu Ghurob, Abusir, Saqqara and Dahshur.
migration Seasonal (generally) movement of a population in response to breeding or feeding stimuli.
morph One of several distinct types of plumage in the same bird species.
Mut Mother goddess and consort of Amun, a chief god of Egypt. She was associated with vultures.
nape Back of neck.
nocturnal Active at night.
nominate Refers to first subspecies to be formally named.
non-passerine All orders of birds except for passerines.
nose-leaf Fleshy embellishment on nose of certain bat species.
order Group of related families.
ostracon (pl. **ostraca**) Limestone or pottery fragment used to sketch, make notes or practise writing.
passerine perching birds, including all songbirds.
Palearctic Old World and Arctic Zone.
pectoral Necklace with a pendant worn so that it hangs over the chest.
pelagic Ocean going.
pied Black and white.
polygamous In breeding, system where one individual has more than one mate.
predator Animal that preys on other live animals.
primaries Outer flight feathers in wing.
primary projection On resting bird, projection of primaries beyond tertials.
pupa In insects, stage (usually static) between larva and adult.
Pyramid Texts Texts inscribed in the burial chambers of pyramids from the end of the Fifth Dynasty until the First Intermediate Period, after which they were usurped by the elite and abandoned by royalty.

race Subspecies.
range Geographical area or areas inhabited by a species.
raptors Birds of prey and vultures, excluding owls.
reptiles Vertebrates traditionally of the order Reptilia, represented in Egypt by the lizards, snakes, turtles and tortoises, and crocodilians.
riverine Of freshwater habitat along rivers.
ruminant Mammal (often a large herbivore) adapted to eating food twice, the second time after regurgitation and 'chewing the cud'.
scapulars Feathers along edge of mantle.
secondaries Flight feathers between primaries and tertials.
spangles Distinctive white or shimmering spots in plumage.
species (sp.) Fundamental taxonomic division. Groups of animals or plants reproductively isolated from other such groups. In practice, this is becoming increasingly clouded.
speculum Area of colour on secondary feathers of wings, for example in ducks.
streamers Long extensions to feathers, usually of tail.
subspecies (ssp.) Distinct form that does not have specific status.
supercilium Streak above eye.
talons Strong, sharp claws used to seize or kill prey.
tarsus Lower part of leg.
taxonomy/taxonomics Science that classifies all living organisms.
terminal band Broad band on tip of feather or tail.
terminal tuft In mammals, tuft of elongated hairs at tips of many species' tails that is often boldly marked in black and white.
territory Area that is defended by an animal or by a group of animals.
tertials Innermost wing-coverts, often covering secondaries.
Thebes The Greek name for ancient Waset, modern Luxor, Thebes was a district capital that rose to prominence when governors became kings of all Egypt at the end of the Eleventh Dynasty. It was greatly expanded during the New Kingdom, when Theban pharaohs again ruled Egypt. Western Thebes was a vast cemetery for the residents of the town as well as for royalty buried in the Valley of the Kings and the Valley of the Queens.
thorax In insects, one of the three body divisions between the head and abdomen bearing legs and wings.
trousers Feathered legs of large birds of prey such as *Aquila* eagles.
underparts Undersurface of a bird from throat to vent.
undertail-coverts Plumage covering base of tail beyond vent in birds.
underwing Undersurface of a bird's wing including the linings and flight feathers.
underwing-coverts Plumage covering underside bases of primaries and secondaries of birds.

ungulates Hoofed mammals belonging to several orders – in Egypt, the Artiodactyla, Perissodactyla and Hyracoidea.
Upper Egypt Area south of Cairo, thus called as the Nile flows from south to north.
upperparts Upper surface of a bird including wings, back and tail.
uppertail-coverts Plumage covering base of upper tail below rump.
upperwing-coverts Plumage covering upperside bases of primaries and secondaries.
uraeus Symbol worn on front of a headdress, representing the mythical fire-spitting cobra, a protector of kings and gods, rearing up with its hood outstretched.
vagrant Accidental, irregular.
vent Undertail area.
ventral Referring to underparts.
vertebrate In taxonomic terms, any member of the subphylum Vertebrata of the phylum Chordata. Broadly an animal with a backbone.
victual mummy Prepared foods ready for consumption, such as joints of meat and poultry (ducks, geese, pigeons) that are preserved and wrapped as mummies, often provided with coffins, and placed in tombs to provision the dead in the Afterworld.
vulpine Of foxes or fox-like.
wadi Seasonal or dry desert valley.
Weighing of the Heart After death the Egyptians believed that they were taken to the Judgement Hall of Osiris, where their heart was weighed against the feather of truth/order/justice (Maat). If the heart and the feather balanced, Osiris would receive them in the Afterlife. If the heart was heavier than the feather, they would be devoured by Ammit, a monster made up of a crocodile, hippopotamus and lion.
wing-coverts Small feathers on wing at bases of primaries and secondaries in birds.
wingspan Length from one wing-tip to the other when fully extended.
winter plumage Bird plumage seen during non-breeding winter months.
Western Palearctic Faunal region encompassing Europe east to the Urals, North Africa and much of the Middle East.

USING THIS BOOK

Each of the 62 animal species described in this book, selected for its religious, cultural or economic significance in ancient Egypt, has its own entry as a living animal accompanied by a photograph, complemented by a parallel entry describing the significance of the animal in ancient Egypt, together with photographs of its depiction in tombs and temples.

LIVING ANIMALS

The photograph of the live animal shows the species in nature. Certain species, such as the Addax and Scimitar-horned Oryx, are now so rare or close to extinct in the wild that photographs of captive animals have been used. The species order follows generally accepted scientific protocol – for instance for birds that of the latest official Egyptian Ornithological Rarities Committee (EORC) of 2019. The species descriptions of the living animals are organized as follows.

Species Name & Measurements The species name is given – both the common English name and the scientific name. The English name in bold is the most widely accepted name across the literature. However, names change and are constantly changing. Where there are widely used alternative names these are given in parentheses and not in bold. Thus the Hamadryas Baboon is also listed by its alternative name of Sacred Baboon. The scientific name in italics is the name generally accepted in the scientific literature. However, taxonomy is also constantly changing as advances are made in genetic studies, and occasionally there are differences associated with such changes. For instance, the African Golden Wolf *Canis lupaster* in the current book follows the latest edition of Castelló, but even in recent literature appears as the Common or Golden Jackal *C. aureus*. This taxonomic discrepancy has little impact on identification for the observer in the field (nor in temple friezes).

The length of the animal is then given. For mammals this is head and body length (HBL) and tail length (TL) as well as shoulder height (SH) for larger mammals and forearm measurement for bats. Weight (WT) is also given. For birds the total length is provided (height (HT) in the case of the Ostrich), as well as wingspan (WS) for larger species. Total length is given for reptiles, amphibians, fish and most invertebrates. The wingspan is given for butterflies.

Description A short general description is followed by overall details using the descriptors detailed in the glossary (p. 24).

Distribution Following the description of each species, its distribution is given within Egypt using the geographical areas described in the section on biogeography. The status within Egypt is then given, again using the geographical areas described earlier. For birds, 'resident' means that the species is present throughout the year

and breeds. 'Summer breeder' indicates that the species breeds but only as a summer visitor. 'Winter' means that the species does not breed and occurs only during winter. 'Passage' indicates that the species occurs on migration in spring and autumn. These descriptions are not mutually exclusive – a species may occur on migration but some birds may also stay and overwinter. Some species are very local, while others may be more widespread, and on migration a species may be encountered almost anywhere.
Habitat & Habits The key habitats in which a species is likely to be encountered are detailed, although note the above point about migratory species. Where space permits, key behavioural details may be highlighted or any distinctive traits; for birds, for example tail fanning, wagging or cocking, distinctive flight patterns, associated species or anything else that may aid identification.
Similar Species Inevitably, not all species recorded in modern Egypt were portrayed in ancient Egypt. The modern visitor will see many other species and similar and related species, and the most significant are covered in these notes. Conversely, many species portrayed in ancient Egypt are no longer found in modern Egypt. When this is the case, their closest relatives are noted.

ANCIENT ATTESTATIONS

The species descriptions of the ancient animals are organized as follows:
Name Common name of the animal followed by the hieroglyphs and the animal's name in ancient Egyptian.
Introduction Brief general introduction to the significance of the animal to the ancient Egyptians, in both religious and practical terms. Similar species that share the iconography of the main species are also mentioned here.
Practical Uses What the animal might have been used for.
Mythology/Iconography The role, if any, of the animal and its associated creatures in mythology/iconography.
Highlights Where viewers might see images of these animals, whether in museums, temples or on tomb walls. For help with the location of each place, a reference is given for each (such as 'TT', for 'Theban Tomb', followed by a number).

Long-eared Hedgehog ■ *Hemiechinus auritus* HBL 136–206mm TL 15–39mm WT 400–500g

A. Main

DESCRIPTION Small, pale, slim-limbed hedgehog with large ears. Upperparts covered with rather short spines reaching 1.5cm along back. Spines darkish-brown at base, creamy with white tips giving overall pale appearance, darker on the North Coast west of Delta (subspecies *H. a. libycus*). Underparts unspined and white tinged yellow-buff. Legs long and slender and clear when trotting, whitish. Feet white. Facial region uniform pale brown with long, pointed snout. Ears large, pale and translucent, with rounded tips; they clearly emerge from the spines hence name. Tail short, indistinct. **DISTRIBUTION** In Egypt restricted to north. Found across North Coast from Salum east to Port Said and Delta, Fayoum and Lower Nile Valley. A few records from North Sinai around Wadi El Arish. **HABITAT AND HABITS** Not a species of true desert but of desert margins, coastal desert, farmland, barley fields, gardens and around village settlements. Locally common, although a frequent roadkill victim and also captured for pet trade. Largely nocturnal, living in a self-excavated burrow. Omnivorous, eating mainly large insects but also fruits and small vertebrates. Probably significant in pest control. Preyed on by large raptors such as the Pharaoh Eagle Owl *Bubo ascalaphus* and mammalian predators. Does not roll up as a ball of spines like typical hedgehogs. Can be a pest in grain stores. Gestation 22 days, with litter size of up to 10. Breeds throughout the year. **SIMILAR SPECIES** The Ethiopian Hedgehog *Paraechinus aethiopicus*, sometimes split into three separate species in Egypt, is found in three distinct populations along the North Coast, in South Sinai and the very south of the Eastern Desert. It is larger, with proportionally smaller ears, bicoloured or dark face and frontal spine with a 'parting'. The Crested Porcupine *Hystrix cristatus* is very much larger.

Hedgehog *Henta*

While hedgehogs were once very common in Egypt, they are less so now due to habitat loss. They are sometimes kept as pets and also used in medico-magical practices. **PRACTICAL USES** As food (covered in clay and baked in a fire) and as materia medica – dried hedgehogs are sometimes found in folk-medicine shops even today. They might have been kept as pets in antiquity. **MYTHOLOGY/ICONOGRAPHY** While no deities are directly associated with hedgehogs, they seem to have been apotropaic with their heads being shown on ships' prows, notably in the Old Kingdom. There are some amulets in the shape of hedgehogs – possibly the idea was that the spines would magically protect the wearer. Some faience vessels are made in the shapes of hedgehogs. They are also shown as populating the landscape beyond the Nile. **HIGHLIGHTS** Found in some Old Kingdom tombs, notably that of Mereruka, where there are two examples of hedgehogs being taken in small cages as offerings. They are also shown as part of the desert landscape where hunting takes place, such as that in the tomb of Mereruka and the causeway of Unas. They feature in Theban tombs that contain hunting scenes. Museums contain figurines and amulets in the form of hedgehogs.

A. Oliveira

A group of hedgehogs; Tomb of Kagemni, Saqqara

S. Ikram

A hedgehog being brought as an offering; Tomb of Mereruka, Saqqara

African Giant Shrew ■ *Crocidura olivieri* HBL 106–135mm TL 57–84mm WT 40g (Greater Musk Shrew, Giant Musk Shrew)

DESCRIPTION Shrews are superficially mouse-like mammals but are insectivores rather than rodents. They are generally small to very small mammals that are highly active, feeding on a variety of invertebrate prey. The African Giant Shrew is a large, rather dark, long-tailed shrew, mouse-like but with elongated, pointed snout, small eyes and furred tail. Fur short, soft and dense. Upperparts uniform dark brown, glossy on flanks. Underparts dark greyish. Head has typically tapered snout. Ears large and sparsely haired. Teeth white. Tail uniform dark brown with dark bristles, especially towards base. **DISTRIBUTION** In Egypt, the most widespread shrew. Recorded from the Delta and the Valley south to Dahshur and also Fayoum. Elsewhere distributed over much of sub-Saharan Africa south to northern Botswana and Zimbabwe. **HABITAT AND HABITS** Areas of moist, dense vegetation such as canal and river banks, irrigated fields and other agricultural areas; also villages and gardens in Fayoum. Solitary. Active by day and night. Builds a nest of damp vegetation such as matted grass. Feeds largely on insects, snails and other invertebrates, but may also take frogs. **SIMILAR SPECIES** The African Giant Shrew was first described from a specimen from Saqqara, 'near Giza'. Today at least four other shrew species are recognized in Egypt, including the House Shrew *Suncus murinus*. The Egyptian Pygmy Shrew *C. religiosa* has been recorded from the Delta and North Sinai. It was originally described from a mummified shrew specimen from Thebes. Flower's Shrew *C. floweri* is an enigmatic species described from a specimen in Giza and rarely seen since. It is one of very few mammals that is endemic to Egypt. Categorized by the IUCN as endangered, it may already be extinct.

Nick Greaves/Alamy

Giant Shrew *Amamu*

A variety of shrew existed in ancient Egypt and these shrews do not seem to be particularly differentiated in terms of nomenclature or symbolism. **PRACTICAL USES** In medicine and magic, and as mummified offerings. **MYTHOLOGY/ICONOGRAPHY** Shrews were associated with the sun god, in particular the nocturnal aspect, as it was thought that they could see in the dark. Actually, shrews have poor eyesight but an extraordinary sense of smell that allows them to navigate regardless of the amount of light. Thus, they became the sun god's creature when he made his journey through the Underworld and were mummified as offerings to him. **HIGHLIGHTS** Thousands of mummified shrews were given as offerings. Many of the mummies were placed in small bronze, wood or stone coffins surmounted by an image of a shrew, which was often gilded to evoke the sun.

S. Ikram

Shrew mummy (6465); (TT 11), Djehuty Project

A. Oliveira

Bronze figure of a shrew; Egyptian Museum (JE 7201)

Bats ■ Order Chiroptera. Megachiroptera **Egyptian Fruit Bat** *Rousettus egyptiacus* Forearm 84–99mm WT 130g Microchiroptera such as **Egyptian Slit-faced Bat** *Nycteris thebaica* Forearm 43–49mm WT 8–11.5g

CC Lietuvos zoologijos sodas

DESCRIPTION Bats are the only mammals capable of true flight, as opposed to gliding, on wings consisting of a skin membrane over the greatly modified forelimbs. The Megachiroptera are represented in Egypt by a single species, the Egyptian Fruit Bat, distinguished from all other Egyptian bats by its much larger size (wingspan up to 60cm) and prominent eyes. The Microchiroptera, the insect-eating bats, are much smaller and are represented in Egypt by 21 species. The order includes a number of different families. Free-tailed bats are large with robust tails and large ears. Rat-tailed bats are smaller with very long, skinny tails. Horseshoe bats have a complex nose-leaf, the shape of which is reflected in the name. Vesper bats are perhaps best represented by Kuhl's Pipistrelle *Pipistrellus kuhlii*, which lacks both an obvious tail and ornate nose-leaf and weighs in at just 5–8.5g. **DISTRIBUTION** The Egyptian Fruit Bat is distributed throughout the southern Delta and much of the Valley south to Aswan, with records also from Port Said, and Suez and Dakhla oasis. The Microchiroptera are found throughout, but absent from true desert. Kuhl's Pipistrelle is common throughout the Delta, including Cairo and south possibly to Luxor. **HABITAT AND HABITS** Fruit bats are generally found in agricultural areas and villages, towns and cities where there are fruiting trees, including Cairo. The insectivorous bats are found throughout and often roost in archaeological sites. At any temple or tomb, they may be detected by a musty odour and by droppings on the ground. All are crepuscular or nocturnal, emerging in the evening to hawk for insects, of which they are a valuable control. **SIMILAR SPECIES** Bats are the only flying mammals. Their flight is generally more fluttering and erratic than that of birds.

Bat Fruit bat *Sakhemu* Insect-eating bat *Dergyt*

While bats have been omnipresent throughout Egypt, particularly in tombs, they were rarely depicted in Egyptian art. Images of both fruit- and insect-eating bats appear in the tomb of Baqet III (tomb 15), at Beni Hasan where they are grouped together with birds as they could all fly. **PRACTICAL USES** Bats may have been used in medicine and magic, and their guano could have been used as fertilizer. **MYTHOLOGY/ICONOGRAPHY** Bats do not feature in Egyptian religion, but are found depicted in tombs at Beni Hasan. **HIGHLIGHTS** A row of bats is depicted, grouped with birds as creatures that fly, at Beni Hasan.

S. Ikram

Bat; Tomb of Baqet III (No. 15)

Hamadryas Baboon ■ *Papio hamadryas* HBL 50–65cm (F) 70–95cm (M) TL 40–60cm WT 10–15kg (F) 15–20kg (M) (Sacred Baboon)

DESCRIPTION Large, robust monkey showing pronounced sexual dimorphism. Mature, sexually active males are grizzled grey-brown with a large cape of pale grey fur covering the forequarters, shoulders and mantle, with whitish cheeks-ruffs. Naked face and rear callosities flesh-coloured to scarlet in breeding males. Eyes deep-set. Muzzle large and doglike, with canines greatly enlarged, especially in males. Tail moderately long, typically arched and with a whiter tip in male. Non-breeding males and females browner and greyer, with flesh-coloured callosities; hands and feet darker. Infants dark brown to black. Vocal, especially in deterring predators and in rivalry within the troop, with loud barks. **DISTRIBUTION** Extinct in modern Egypt. Elsewhere found in the Horn of Africa from eastern Ethiopia to Sudan and Djibouti to the Somali border; also in the southwestern Arabian Peninsula in Saudi Arabia and Yemen. **HABITAT AND HABITS** Found in mountainous regions along the Red Sea and in arid, semi-desert areas along escarpments, hillsides and wadis. Lives in troops generally with one mature adult male and a harem of females, youngsters and infants. Spends a lot of time grooming. Larger groups, known as bands and clans, also occur. The Hamadryas Baboon is an omnivore, feeding on a variety of wild fruits and grasses, but also refuse from dumps and abandoned camps. Expanding agriculture has the potential of human–baboon conflict. **SIMILAR SPECIES** Other baboon species have been recorded from mummified remains dating back to Predynastic times. The two cited are the Olive Baboon *P. anubis* and Yellow Baboon *P. cynocephalus*. Neither of these species occurs in modern Egypt. There are extensive illustrations of these species in John Anderson's *Zoology of Egypt: Mammalia* (1902).

Hamadryas Baboon *Iana*

S. Ikram

Baboon; Tomb of Tutankhamun (KV 62)

S. Ikram

Baboon; Tomb of Sennedjem (TT 1)

Although baboons do not seem to have ever been native to Egypt, they appear in Egyptian art from at least 3100 BC in the form of statuettes, possibly representing an obscure deity, the 'Great White'. The zooarchaeological remains of baboons have been found at the site of Hierakonpolis, a major centre of Predynastic Egypt, but these animals seem to have been imported from further south and kept in captivity. The many baboon mummies found dating from 600 BC onwards also seem to be derived from imported populations that may have been bred in temples. **PRACTICAL USES** Baboons appear to have been used by the police to catch thieves, as illustrated in some Old Kingdom tombs and temple complexes, as helpers in boat-building and in collecting fruits. The bones of baboons have been found in palaces, suggesting that they had served as guardians, helping to patrol the premises. They were also mummified as offerings to Thoth and Re, with their main cult centres being at Saqqara and Tuna el-Gebel. Some baboons were also kept as pets. **MYTHOLOGY/ICONOGRAPHY** The baboon was sacred to both the sun and the moon, and thus associated with Re and Thoth, respectively. The solar association of the baboon might have been due to the fact that its face is reddish and its ruff fans out and evokes the sun; more obviously, its posterior is bright red, again evoking the sun. Another reason for the baboon's solar association might be the fact that on waking, baboons face the rising sun, stretch and emit grunting cries, suggesting that the baboon is greeting and aiding the rising sun – they are shown thus at the bases of obelisks, a quintessential solar symbol. Of course, this would indicate that the Egyptians saw this animal in nature. The lunar association of the baboon is more difficult to understand as it is a diurnal creature. However, it is possible that its silvery coat evoked the moon (the 'Great White' refers to its silvery-white coat). Other baboon deities include Babi, the 'Bull of the Baboons', who consumed the entrails of the unrighteous dead, but bestowed virility upon the righteous. Hapi, one of the Four Sons of Horus, who guarded the lungs after their removal during mummification, was baboon headed. **HIGHLIGHTS** Two- and three-dimensional images of baboons abound in Egyptian museums and museums worldwide. The Saqqara and Giza mastabas show baboons at work, as do some tombs at Beni Hasan, and they feature prominently in religious contexts in the royal tombs of the Valley of the Kings and in funerary (and other) temples in the Luxor area, as well as at Abu Simbel, where they are located in areas that are the first to receive the rays of the sun. Giant baboons are located at the site of Ashmunein at the temple to Thoth.

Grivet Monkey ■ *Chlorocebus* (Formerly *Cercopithecus*) *aethiops* HBL 38–61cm (F) 45–83cm (M) TL 42–114cm WT 2.5–6kg (F) 3.5–9 kg (M) (Vervet Monkey, Green Monkey, Tantalus)

Shutterstock/R. Harding

DESCRIPTION Medium-small, long-tailed, archetypical African monkey. Adult grizzled grey to greenish-brown above, paler below, even to white. Hair coarse. Face has black facial mask with white across forehead and down sides of cheeks. Hands and feet pale. Thumb very small. Tail long, tipped with white, with no prominent terminal tuft. Scrotum blue to bright blue. Variable and with taxonomy constantly changing. *C. aethiops* is the northernmost of the savannah monkey complex. **DISTRIBUTION** Extinct in modern Egypt. Elsewhere, it ranges across South Sudan and Ethiopia and Eritrea north along the White Nile to Sudan. The wider genus is distributed over much of sub-Saharan Africa south to South Africa. **HABITAT AND HABITS** Open woodland, savannah, riverine, cultivated and agricultural habitats. Largely terrestrial, living in mixed troops of up to 70 individuals with an established hierarchy. Needs access to water, so absent from desert habitats. Particularly dependent on *Acacia* species, but feeds on a wide variety of fruits, leaves and seeds, as well as invertebrates and other animal food. Largely diurnal, retreating to trees or cliffs at night. While widespread and common, in populated agricultural zones, it comes into conflict with farmers. Also exploited as part of the pet trade. **SIMILAR SPECIES** Baboon species much larger, more uniform and without black face. Within historicalal times the Common Patas Monkey *Erythrocebus patas* may have occurred in Egypt. Its current range extends across the Sahel including Sudan, where it is found in dry woodland and open savannah habitats. It is much more rufous and russet than the Grivet, and more long limbed and slender. It is renowned as the fastest runner among the primates, reaching up to 55km/ph.

Grivet Monkey *Key* or *Gef*

Grivet monkeys as well as a few other species were imported into Egypt, probably as companion or status animals, from the mid-fourth millennium onwards. **PRACTICAL USES** Companion animals, possibly to help with fruit gathering. **MYTHOLOGY/ ICONOGRAPHY** The Green Monkey has no particular role in mythology, although it was possibly associated with the sun god. It was thought to be a playful and sometimes mischievous creature and is featured on ostraca and papyri as such. The determinative for monkey takes the form of a baboon, indicating that the baboon was regarded as the most typical and most important monkey. **HIGHLIGHTS** These monkeys appear in the tomb of Nefermaat and Itet (Meidum) now in the Egyptian Museum, as tribute in Rekhmire's Theban Tomb (TT100) and that of Amunedjeh (TT84), and with other monkeys in the tomb of Baqet III at Beni Hasan, and in Behera's tomb at el-Bersha, as well as on many ostraca. Kohl and other eye-cosmetic containers take the form of a monkey or a mother and baby monkeys. Mummies of pet Grivet Monkeys have been found in tombs such as that of Tia and Tia at Saqqara, and they also feature as mummified offerings in the catacombs at Tuna el-Gebel and Saqqara.

S. Ikram

A Green Monkey crouched under its owner's chair eyeing a pile of Nabk berries; Tomb of Userhat (TT56)

S. Ikram

Two pet monkeys frolicking under their owner's chair and devouring grapes; Tomb of Ptahemwia, Saqqara

Red Fox ▪ *Vulpes vulpes* HBL 46–65cm TL 30–40cm SH 35–40cm WT 1.8–3.8kg

R. N. Hewison

DESCRIPTION Slim, dog-like carnivore with a very full, bushy tail. Largest Egyptian fox with female on average slightly smaller than male. Coat greyish-red to grey-brown above, darker on nape. Flanks greyish with buff tinge. Throat and belly dark, even blackish. Chin white. Legs brownish with black stripe down rear side on forelegs. Slender, vulpine muzzle. Ears large with back of ear black. Desert foxes tend to be paler than those from, for example, the Delta. Red Foxes look shabby in early spring as they lose their winter coat. Vocal, especially during rut. Variety of barks, growls and whines. **DISTRIBUTION** In Egypt recorded from throughout the Delta and across the north coast from Alexandria west. Also south and west Sinai, along the northern Red Sea coast and around Lake Nasser. In the Western Desert from Kharga, Dakhla and Fayoum. Elsewhere over much of Europe, North Africa, and across Asia including the Middle East. **HABITAT AND HABITS** Farmland, gardens, scrub and desert margins. Can readily be seen at tourist sites such as the Step Pyramid at Saqqara. Range seems to be expanding with tourist development, especially along coasts. Mainly nocturnal, generally holing up in a burrow or earth by day. Omnivorous and opportunistic. Feeds on a wide variety of vegetable and animal foods, including fruits such as melons, eggs, large insects, fish, small mammals and birds. Excellent hearing, scent and sight. **SIMILAR SPECIES** Three other fox species are found in Egypt, the most likely of which to be encountered is the smaller Rüppell's Sand Fox *V. rueppelli* which has buff backs to the ears. The Fennec Fox *V. zerda* is much smaller, uniform cream, with huge ears and a black tip to tail. Blanford's Fox *V. cana* is very rare in South Sinai and the Eastern Desert, with a huge tail and distinctive black from below the eye to the muzzle.

Red Fox

Possibly *Sab*, also used for jackal; *Meset* is 'Apron of Fox Skins'

Foxes mainly appear in a limited number of scenes of the desert and hunting. **PRACTICAL USES** The fur, especially the tails, could be used as clothing or decorative elements in regalia; an apron of fox skins is mentioned in texts. Parts of fox were also used for religious, magical or medical purposes. **MYTHOLOGY/ICONOGRAPHY** Foxes could be conflated with Anubis, god of embalming and travel in this world and the next, as foxes were expert navigators of the deserts. Fox mummies (foxes of different species) are found in catacombs dedicated to Anubis in Saqqara, as well as at Asyut. For some reason, foxes or fox skins were associated with birth: the sign *mes*, part of the word meaning 'to be born', consists of three foxes (or fox skins) connected at their muzzles. Hence the so-called birth-wands or clappers, generally made of Hippopotamus ivory, that might have been used in rituals associated with birth, both human and divine, as well as rebirth, and had fox heads carved on their tips. **HIGHLIGHTS** Predynastic objects sometimes depict foxes, as well as men wearing fox tails (as seen on the Hunter's Palette now divided between the British Museum and the Louvre). Foxes are found in hunt scenes in the Memphite mastabas and Theban tombs, including that of Rekhmire (TT100), a tomb that also features fox skins being brought as tribute from Nubia. They appear at the ends of clappers/wands (in both the Grand Egyptian Museum and Egyptian Museum), and on ostraca in several museums in Egypt and worldwide.

S. Ikram

A red fox, under the horse, lies wounded in a hunt scene; Tomb of Userhat (TT56), Luxor

African Golden Wolf ■ *Canis lupaster* HBL 72–93cm TL 29–35cm SH 38–40cm WT 10–15kg (African Wolf, formerly Common Jackal, Golden Jackal)

Mourad Harzallah

DESCRIPTION Medium-sized, dog-like carnivore best described as a small and slightly shaggy wolf and now treated as such. Typical dog-like form with a coarse coat of shaggy, grizzled fur looking greyish-brown at any distance. Paler below and on insides of legs and throat. Legs buffish with black along back of foreleg. Mane of longer hair along back. Muzzle slender. Ears relatively small and pointed, rufous behind. Tail rather bushy and black along top and tip. Yelps and barks when excited, and growls. **DISTRIBUTION** In Egypt recorded from the Western Desert including the Western Oases, the Delta and environs; also Upper Egypt, including shores of Lake Nasser and locally in North Sinai. Elsewhere over much of North Africa and the Sahel south to the Horn of Africa. Widespread but probably declining where competing with feral dogs. **HABITAT AND HABITS** Agricultural areas, scrub and desert margins. Rocky habitat and lake shore on Lake Nasser. Not a desert animal except for the semi-arid Northern Coastal Strip. Largely nocturnal but can be active at dusk and by day. Omnivorous and opportunistic, feeding on a wide variety of vegetable and animal sources, including melons, corn, carrion, small mammals, birds, insects and fish. Excellent hearing and scent, and good sight. **SIMILAR SPECIES AND NOTES** This is the 'jackal' with which Anubis is most commonly associated. Formerly known as the Common or Golden Jackal, it is now widely regarded as a separate species from the Jackal C. *aureus* found over much of southern Europe, the Middle East, and South and Southeast Asia. Egyptian fox species are much smaller and slimmer.

Jackal/Wolf — or *Wenesh*

Images of African Goldern Wolves abound in Egyptian art. What were identified as jackals by the modern and probably the ancient inhabitants of Egypt, as well as scholars, are now thought to be a form of wolf (opposite) based on genetic analyses. However, the animals look very jackal-like – certainly the ancient Egyptians would have identified the animal phenotypically rather than genetically. **PRACTICAL USES** None known, save for medicine and magic. The tails could be used as an accoutrement. **MYTHOLOGY/ICONOGRAPHY** Jackals were associated with the gods Wepwawet and Anubis. Wepwawet means Opener-of-the-Ways, as jackals create paths in the desert and can navigate both the Nile Valley and the desert. Wepwawet was initially a war-like deity opening the way for the king's victory. He also opened the way for the transition to the afterlife, a role that was subsequently taken over by Anubis. Wepwawet can be differentiated from Anubis (when colour is preserved) as his coat is grey while that of Anubis is black, or by the hieroglyphs labelling him. It has been posited that Anubis is actually not only a jackal, but an amalgam of jackal, fox, dog, possibly wolf, and hyena (not a canine, but resembling one), creating a super canid. Often these two gods are paired and shown flanking Osiris. In addition to taking on Wepwawet's role as guide to the Afterworld, where the weighing of the heart took place, Anubis also officiated at mummification and was the patron god of the embalmers. Presumably this association with death was due to the fact that jackals and other canines would haunt cemeteries and embalming areas to make off with carrion, and in order to protect against this the Egyptians sought to connect these animals with a god who would guard the dead against such creatures. **HIGHLIGHTS** Anubis is represented in either jackal form or as a man with a jackal's head. Images of Anubis abound in two- and three-dimensional art in museums and on tomb and temple walls. He appears in many scenes in the Valley of the Kings as well as in the tombs of the Nobles in Luxor. In Saqqara he is pictured in some of the New Kingdom tombs, such as the subterranean section of the tomb of Horemheb.

S. Ikram

Anubis as a super-canid, mainly a jackal; Tomb of Roy (TT 255)

S. Ikram

Anubis holding the ankh, the key of life; Tomb of Thutmose IV (KV 43)

Domestic Dog ■ *Canis lupus familiaris* or *C. familiaris* HBL Variable T Variable WT Variable

Richard Hoath

DESCRIPTION The Domestic Dog has been domesticated longer than any other animal species, dating back at least 15,000 years and possibly as long ago as 30,000 years. Genetic studies suggest that domestication began around 23,000 years ago from a population of an ancestor of the Eurasian Grey Wolf *C. lupus*. Today, hundreds of different breeds of the Domestic Dog are recognized, which can be classified as four major types – Old World Dogs, Mastiff type, herding type and hunting type. Representations of the Domestic Dog in ancient Egypt date back to Predynastic times, and a variety of the above types can be recognized. **DISTRIBUTION** In Egypt, feral dogs are a familiar sight throughout. **HABITAT AND HABITS** Egypt in virtually every human habitat, from desert margins to city centres, often in small packs. **SIMILAR SPECIES** The Arabian Wolf *C. l. arabs* HBL 70–80cm TL 30–34cm SH 40-65cm WT 18–20kg, at up to 20kg, is one of the smallest races of the widespread Eurasian Grey Wolf *C. lupus*. Similar to the African Golden Wolf *C. lupaster* but longer legged with a shorter, less shaggy coat, largely light brown to yellowish-grey and with a crest running from the base of the head to the base of the tail. Tail with dark tip. In Egypt recorded only from Sinai, where either transient or at best very rare. Elsewhere across the Arabian Peninsula north to Syria, where replaced by the Indian Wolf *C. l. pallipes* but everywhere rare and threatened. It lives in desert and semi-desert mountains and plains, but requires water. Solitary, in pairs or small groups; only in larger packs where there is sufficient prey. Feeds on mammals up to the size of gazelles, birds and reptiles, and also carrion, invertebrates and fruits.

Domestic Dog *Iwiw* or *Chesem* (hunting hound)

The dog is the first animal to have ever been domesticated anywhere in the world and features in Egyptian art from prehistory through the end of the pharaonic Era. Its name in Egyptian, *iwiw*, is onomatopoeic. Different types of dog are depicted in Egyptian art, with many that are of the saluki-type, and others with short limbs, although the majority of those found are similar to the 'baladi' or local dogs found in the streets of Egyptian cities as well as at archaeological sites throughout the country. **PRACTICAL USES** Dogs have been used as guardians, as hunting and herding aids, as companion animals, and as offerings to canine deities, notably Anubis. They could also be used in medicines and as part of magical rituals. **MYTHOLOGY/ICONOGRAPHY** Dogs and other canines were associated primarily with Wepwawet and Anubis. Wepwawet is an ancient deity, perhaps the oldest canine deity, associated with what are now called African Golden Wolves, but were formerly identified as jackals. In the later periods of Egyptian history, offerings of mummified dogs were consecrated to Anubis, and some cemeteries, such as the one associated with the Anubeion at Saqqara, once contained as many as eight million dog mummies that ranged from Maltese types to a few that were the size of mastiffs. Dogs are frequently shown under the chairs of their owners, following their owners or in hunt scenes. **HIGHLIGHTS** Dogs appear in Egyptian art from the Predynastic Period onwards, and are depicted hunting on the objects from the tomb of Hemaka, in many mastaba tombs of the elite at Saqqara (such as those of Mereruka and Ptahhotep) and Thebes, and under the chairs of their masters or mistresses in tombs in Saqqara and Thebes. The tomb of Sirenput I at Qubbat el-Hawa shows different types of dog following their owner, and several dogs also feature in the tombs at Beni Hasan, ranging from hunting hounds to small, squat, sausage-shaped dogs. A wooden coffin from Asyut in the Egyptian Museum shows a man walking his spotted dog on a leash.

J. M. Galán

Chesem hounds; Mastaba of Mereruka, Saqqara

S. Ikram

Dogs; Tomb of Sirenput, Kubet el Hawa

African Wild Dog ■ *Lycaon pictus* HBL 80–110cm TL 30–40cm SH 65–80cm WT 18–30kg

DESCRIPTION Large canid (largest in Africa), lanky and variably coloured in black, browns, tawny and white. Very distinctive, with strikingly tortoiseshell coat that is patched asymmetrically and differs between individuals, but invariably with a white tip to the bushy tail. Large and rangy with long limbs. Ears broad and rounded. Head has heavy muzzle, generally with a dark line from the muzzle to between the ears. Pups darker than adults. Very vocal species, especially before hunting, with a high-pitched twittering. **DISTRIBUTION** Extinct in modern Egypt and probably from early pharaonic times. Representations from Predynastic times, but unlikely to have survived much beyond the Old Kingdom. Today present in declining numbers over much of sub-Saharan Africa, where everywhere threatened by habitat destruction, persecution and conflict for resources with pastoralists. Races in West Africa, Sahel and from Chad to Somalia now very rare or locally extinct. **HABITAT AND HABITS** Generally in open habitats from semi-desert through savannah to open forests, but has been recorded from closer forest and from montane environments. Lives in packs of half a dozen to tens of animals (previously many more), and has a highly complex social structure with only one female dominant at a time. Hunting is by packs, and prey, most often medium-sized antelopes but also up to the size of the African Buffalo *Syncerus caffer*, is run down over organized and often prolonged hunts. With larger prey the muzzle is targeted, while other pack members bite at other regions. Food is shared by all members of the pack. **SIMILAR SPECIES** None. The African Wild Dog is unique among canids in not having dewclaws on the forelimbs. It is now confined to a few increasingly small and isolated populations, many of which are highly threatened.

CC African Conservation

African Wild Dog No name in Egyptian

No skeletal remains of wild dogs have been found in Egypt, although some are known from Nubia. There is only one artefact that might have had wild dogs depicted on it – a slate palette. **PRACTICAL USES** None. **MYTHOLOGY/ ICONOGRAPHY** If depicted, it is as powerful wild creatures that define the borders of the wilderness and need to be controlled. **HIGHLIGHTS** The only image that might be that of wild dogs is the Two Dog Palette in the Ashmolean Museum in Oxford, England, although several scholars have suggested that the animals depicted are hyenas, common in Egypt, rather than wild dogs, though the ears are more like those of wild dogs.

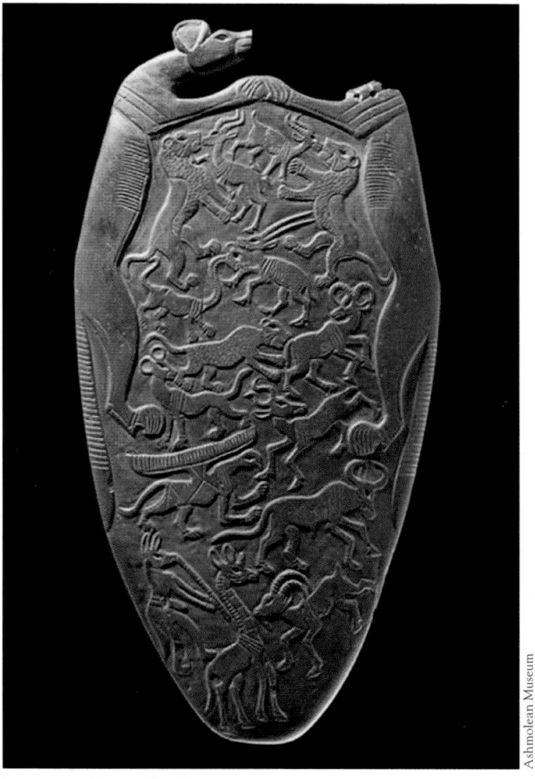

Ashmolean Museum

Two Dog Palette; Ashmolean Museum (AN1896-1908 E.3924)

Eurasian Otter ■ *Lutra lutra* HBL 57–95cm TL 35–45cm WT 7–12kg

DESCRIPTION Large, semi-aquatic mustelid with long, thick-stocked tail. Long bodied and long tailed. Uniform brown above and pale cream below to throat and cheeks. Muzzle blunt and broad with extensive whiskers. Eyes small, set high. Ears small and rounded. Legs rather short, with large feet with extensive webbing. Tail long and tapered, with a broad stock. Fur as body and with no bands or terminal tuft. Vocal, uttering high-pitched twittering. **DISTRIBUTION** Not recorded in modern Egypt. Elsewhere across much of Europe, south to north-west Africa and east across Central Asia to eastern Russia, China south to South-east Asia. Isolated population in southern India. **HABITAT AND HABITS** Tied to water. Unpolluted freshwater habitats such as lakes, reservoirs, rivers and streams. Also occurs on coasts, but requires fresh water for grooming. Decline of many European populations last century largely due to water pollution, as well as active persecution and habitat loss. Has recovered in some areas, for example the UK. Solitary, but female may be seen with cubs. Territorial, with territories marked by spraints (otter dung). Generally nocturnal, resting up by day in a burrow or natural crevice known as a holt. Prey mostly freshwater fish, but also birds, small mammals, amphibians and crustaceans such as crabs. Sole ancient Egyptian representation is of an otter eating a fish in a papyrus habitat. **SIMILAR SPECIES** Only one representation in ancient Egyptian art, for which see opposite. No mummies or skeletal remains. Two other otter species occur in Africa along the Nile. The African (Cape) Clawless Otter *Aonyx capensis* occurs over much of sub-Saharan Africa except the Congo Basin north to southern Sudan and Eritrea. It is larger than the Eurasian Otter, reaching up to 36kg in weight and 160cm in length including tail, and lacks claws on the digits of the front paws. It lives in fresh water with extensive waterside vegetation. The Spotted-necked Otter *Hydrictis maculicollis* has a similar range, although it is also found in the Congo Basin. It too is found along the Nile, from Lake Victoria north to southern Sudan. It is much smaller than the Eurasian Otter at up to 6.5kg in weight, and darker brown throughout with creamy blotches on the throat. It is conceivable that either of these two species could have been depicted in Mereruka's tomb.

Otter No known word in Egyptian

S. Ikram

Detail of the temple carving below

There is a debate as to whether otters ever existed in Egypt. Some bronzes dating to the Late and Ptolemaic Periods depicting animals standing on their hindlegs have been (probably mistakenly) identified as otters. **PRACTICAL USES** None. **MYTHOLOGY/ ICONOGRAPHY** None. **HIGHLIGHTS** The only image that might be that of an otter is found in the tomb of Mereruka at Saqqara. It is in the chamber that contains the statue of Mereruka and can be found on the left side of the door that leads to the chamber. The charming scene shows a creature that looks very much like an otter consuming a fish.

S. Ikram

Possible otter eating fish; Tomb of Mereruka

Egyptian Mongoose ■ *Herpestes ichneumon* HBL 54–61cm TL 36–46cm WT 1.9–4.0kg

CC A. Liflyand

DESCRIPTION Slender, cat-sized carnivore not unlike a large, shaggy weasel. Overall short legged and elongated, with a relatively sharp snout. Upperparts grizzled grey throughout with grizzled appearance, forming dark and pale banding on the long guard hairs. Underparts paler. Legs relatively short and appear more so due to long fur. Feet blackish. Head has rather sharp blackish muzzle, otherwise as upperparts but fur shorter. Ears small and rounded. Tail long, grizzled grey and with blackish tuft of long hairs at tip. Appears wedge shaped, with longest hairs towards tail-base tapering towards tip. Rarely heard but vocal, with chattering, low whine and growls when threatened. **DISTRIBUTION** In Egypt largely restricted to the Delta, including records from the northern lakes, and the Valley to the outskirts of Cairo at Gebel Asfar. Also occurs at Fayoum and eastern parts of the North Coast. Elsewhere southern Iberia, where possibly introduced, Turkey and western Middle East. Much of sub-Saharan Africa. **HABITAT AND HABITS** Mainly farmland and agricultural areas with water. In Fayoum recorded from villages, lake margins, irrigation canals, and so on. Sociable and can be seen in pairs and family groups. Litter size 2–4, and in Egypt there seems to be no fixed breeding season. Lives in a den that may be natural, adopted or self-dug in dense vegetation, reedbeds, rock crevices or among boulders. Wide diet of small mammals, birds, reptiles (including venomous snakes such as cobra species, from which it is protected by its thick fur), frogs and toads, eggs, large insects, fruits and carrion. Senses of sight, hearing and scent all good, with scent perhaps the most important. **SIMILAR SPECIES** Unlikely to be mistaken for any other mammal in Egypt. The much smaller Egyptian Weasel *Mustela nivalis subpalmata*, sometimes treated as a distinct species, is restricted to the Delta south to Cairo and Fayoum. In Cairo it is common even in the centre of the city, where it can be seen at night and early morning scuttling across the streets and dashing beneath parked cars.

Mongoose 𓄿𓂧𓆓 *Adj*

The mongoose was frequently pictured in Egyptian art in both secular and sacred contexts. **PRACTICAL USES** It has been suggested that the mongoose was used as a hunting aid since hunting scenes sometimes show a hunter holding the tail of a mongoose as it stalks its prey of birds, chicks and eggs in the marshes. These animals can be tamed and kept as pets to control vermin, notably rodents, as well as snakes, which were abundant in Egypt. However, it is more common to see the mongoose independently climbing up a papyrus stem to steal eggs or young chicks. **MYTHOLOGY/ICONOGRAPHY** The mongoose (or ichneumon) was associated with several deities: the solar forms of the gods Horus and Atum, and the goddess Mafdet, who protected against venomous animals such as snakes. Statues of these animals as well as mummies were associated with Horus and Atum starting in the Late Period. **HIGHLIGHTS** Images of mongooses are found in the marsh scenes in the tombs at Saqqara. Three-dimensional images of these animals in bronze, dating from the sixth to the first century BC, are found in several museums in Egypt and elsewhere.

S. Ikram

Mongoose climbing a papyrus thicket; Tomb of Idut, Saqqara

Striped Hyena ■ *Hyaena hyaena* HBL 102–110cm TL 29–35cm SH 55–6cm WT 19–45kg

DESCRIPTION Large and dog-like, with shaggy striped coat. Front-heavy carnivore with powerful forequarters. Shaggy coat pale greyish to beige, with a series of narrow, dark vertical stripes breaking into spots on neck and flanks. Distinct mane of long, shaggy fur from neck to base of tail. Head broad with short muzzle greyish, darkening to snout. Ears broad and rounded, reverse black-brown, with inside pale. Throat dark. Legs pale with clear dark bands on outer side, more spotted on inner side. Feet blackish. Tail moderately short, rather full, tapering to pointed tip. Pale beige, darker above with dark tip. Not highly vocal but has a low howl and grunts and whines. **DISTRIBUTION** In Egypt, historically over northern Western Desert, including the Western Oases, and South Sinai. Also occurs in the southern Nile Valley around Lake Nasser and further east. Much depleted everywhere, but still found in Protected Areas. Elsewhere in declining numbers over much of North Africa south to Tanzania and South Asia. **HABITAT AND HABITS** Semi-desert and desert margins, mountains and wadis, stony deserts with *Acacia* species. Previously along camel routes from Sudan. Solitary or in pairs, occupying a large home range marked by scent. Nocturnal, spending much of the day in a den or natural cave or shelter. Wide diet of small mammals, birds, reptiles, eggs, large insects, fruits, and carrion including bones and human rubbish. Hearing and smell acute. Sight good. **SIMILAR SPECIES** The Striped Hyena has been heavily persecuted in Egypt for reported attacks on domestic herds, as well as suffering from habitat loss, reduction in natural prey and increasing expansion of human activity. It has a traditional reputation for savagery and gluttony, yet its body parts have been traded for reputed aphrodisiac effects, for traditional medicine and against the evil eye. However, it is still found in Protected Areas such as Wadi Allaqi and St Catherine's and has been photographed with cubs in the latter. The Striped Hyena should be unmistakable in Egypt, with Common Jackals being much smaller, much less thickset and with shorter, unpatterned coats. The only other hyenid recorded from Egypt is the similarly patterned but much smaller and slimmer Aardwolf *Proteles cristatus*. It is known from two individuals from Gebel Elba.

CC R. Deshmukh

Hyena *Hechet*

Hyenas figure in Egyptian art from the Predynastic to the Late Period after which time they are rarely shown. **PRACTICAL USES** Hyenas were eaten, served as components of medicines and various parts of hyenas were used in magic. **MYTHOLOGY/ICONOGRAPHY** Hyenas were regarded as very fierce, savage creatures. They are among the wild animals that had to be subdued in order to maintain order (*maat*) over chaos and feature prominently in hunt scenes. In the Old Kingdom, hyenas, like many other wild animals, were captured (both adults and cubs) and force-fed before being consumed. **HIGHLIGHTS** The Memphite tombs contain several images of hyenas, adults and cubs, being brought as offerings. There are also some tombs in Saqqara, such as that of Mereruka, in which hyenas are being force-fed delicacies in order to change the flavour of their meat so that they could be eaten. Tombs at Beni Hasan contain images of hyenas being hunted, as do tombs in Thebes, most notably those of Rekhmire (TT100) and Userhat (TT56), and in the hunt scene on Tutankhamun's box (Carter No. 021, JE 61467), hyenas figure among the hunted creatures.

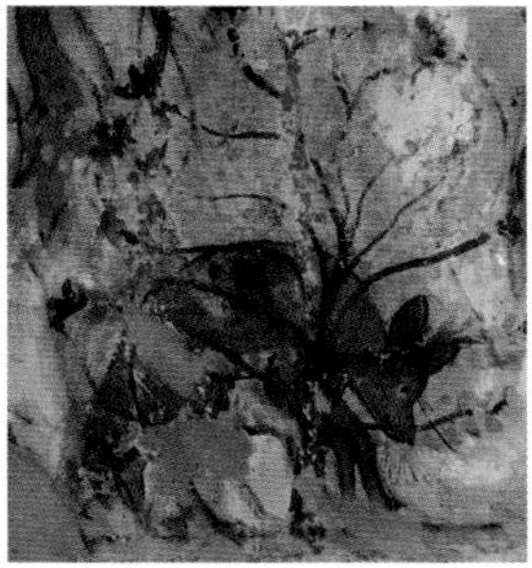

S. Ikram

Hyena caught in a bush; Tomb of Userhat (TT56,) Thebes

A. Oliveira

Standing hyena; Tomb of Mereruka, Saqqara

S. Ikram

Force-feeding hyenas

African Wild Cat ■ *Felis lybica* HBL 38–54cm TL 23–39cm SH 30–36cm WT 1.5–6kg Males generally larger and heavier than females

Shutterstock/Ecoprint

DESCRIPTION Overall appearance just as a Domestic Cat *F. catus*, for which it is the wild ancestor, but generally slightly larger. Typical cat form but longer limbed and with proportionately longer tail. Grizzled buffish above with dark stripe down centre of back and flanks paler. Pale buff to whitish below, marked with pale brown spots that merge to form stripes along shoulder and sides. Brown bands across legs. Muzzle short and white patch below eye. Cheeks striped. Ears with no tuft and diagnostically orange-buff behind. Tail long, blackish above and with dark rings and tip. In Sinai race *F. l. tristrami* darker and more strongly patterned. **DISTRIBUTION** In Egypt historically recorded from the margins of the Delta and Valley. Also occurs at the oases of Bahariya and Dakhla and Fayoum, where once common, and along the North Coast, and parts of Sinai. Elsewhere across much of Africa except the Sahara and Congo rainforest, and across the Middle East through Iran to India. **HABITAT AND HABITS** Dry areas with cover, scrub, desert margins, rocky areas in mountains and cultivated land along North Coast. Habits little known in Egypt, but probably strictly nocturnal. Preys on small mammals, birds, reptiles and arthropods. Solitary, with male maintaining a territory. Agile, and adept climber. **SIMILAR SPECIES** The near ubiquitous Domestic Cat can be very similar in the tabby phase, and perhaps the best distinction is the orange-buff behind the ears. Interbreeding with the Domestic Cat perhaps represents the biggest threat to the African Wild Cat. The Swamp Cat *F. chaus* is a larger, shorter-tailed species recorded from Fayoum and the Valley south to Aswan.

Wild Cat *Miw* (also the word for domestic cat)

DESCRIPTION Wild cats of different species were found in a variety of habitats in Egypt and are shown in swamps as well as in the desert environment. The mythology of different felines overlapped, regardless of their being wild or domesticated. **PRACTICAL USES** To be hunted, in medicine and magic, and as animal mummies offered to the goddess Bastet. **MYTHOLOGY/ ICONOGRAPHY** Cats of all types are seen as little lions and are associated with a variety of deities, primarily those with solar affiliations, including Re, the sun god. Fiercer aspects of cats (and big cats) were associated with the goddesses Mut, Sekhmet and even forms of Hathor, when the cat is protector (Mut, Hathor) or bringer of plague to wrongdoers. **HIGHLIGHTS** Wild cats and swamp cats are shown in different tombs in the Memphite and Theban necropoleis in swamp scenes and other scenes illustrating the natural world. Caracals, too, are shown being hunted. A multitude of cat mummies, and their coffins and sarcophagi, are on display in Egyptian museums as well as museums worldwide. Some bronze and wooden images of cats served as coffins for their mummies, containing wild as well as domestic cats.

S. Ikram

Swamp cats were wild in ancient Egypt; Tomb of Nebamun, British Museum (EA 37980)

S. Ikram

Cat stalking birds in a papyrus swamp; Tomb of Menna (TT69), Luxor

Domestic Cat ■ *Felis catus* HBL 46cm TL 30cm SH 27–25cm WT 4–5kg males generally larger and heavier than females (Feral Cat)

R. Hoath

DESCRIPTION Overall appearance as the African Wild Cat but very variably coloured. The Domestic Cat can be tabby, and very similar to the wild ancestor, but also piebald, marmalade (orange with tabby markings), all white, all black, blue and calico (black, white and orange) or a combination of these colours. Pure-bred domestic breeds now include Siamese, Burmese, Maine Coon and many more breeds but 'street cats' known locally as *baladi* cats, breed back to type just like the Domestic Dog (p. 46). However, while the Domestic Dog has been bred over many thousands of generations to perform a variety of functions including hunting, guarding, herding, companionship, etc., the Domestic Cat has mostly co-existed rather than being deliberately bred and the modern breeds mentioned above have largely been a product of relatively recent intentional selection. **DISTRIBUTION** In Egypt, almost ubiquitous in any populated area from villages to the centre of major urban areas. **HABITAT AND HABITS** Commensal and associated with humans. Domestic Cats can be completely dependent on humans or live as feral cats either individually or in loose colonies where each cat within the colony holds a territory. **SIMILAR SPECIES** The near ubiquitous Domestic Cat can be very similar to the African Wild Cat (p. 56). While the Domestic Cat can be very variable as detailed above there is a distinct breed known as the Egyptian Mau. This is unusual amongst domestic cat breeds in that it has distinctive spots. While it has gained in popularity since the 1950s and is now recognized as a distinct breed, there is no evidence of an actual link to ancient Egypt. Indeed, while cat domestication has popularly been associated with ancient Egypt, it is now thought that cat domestication began elsewhere, possibly the Levant.

Domestic Cat *Miw* (also the word for wild cat)

A. -M. Kellen

Egyptian Museum (CG29776)

In contemporary popular cultures, worldwide, cats are intimately associated with ancient Egypt. Although current DNA research indicates that cats were domesticated in the Levant, they were dispersed throughout the world from Egypt, and have been a crucial part of Egypt's cultural landscape since pharaonic times. **PRACTICAL USES** Controlling vermin, as companion animals, as animal mummies offered to the goddess Bastet. **MYTHOLOGY/ICONOGRAPHY** Like wild cats, domestic cats were associated with solar deities. Domestic cats, in particular, were the totemic animal of the goddess Bastet, in charge of motherhood, beauty and self-indulgence, all attributes of cats. While dogs were given names, cats were generally just called 'cat' or 'the cat', no doubt due to the fact that these animals seldom respond to being called – indeed, they are the owners of their humans! **HIGHLIGHTS** From the Middle Kingdom onwards, particularly in Theban Tombs, pet cats are shown under the chairs of their owners, either seated or devouring choice tit-bits. Sometimes they are shown on their owners' laps. Cats were often offered to the goddess Bastet and buried in the millions at her sacred sites, the best-known ones being Bubastis in the Delta and the Bubasteion at Saqqara. Of particular interest, the Egyptian Museum hosts the sarcophagus of a cat, 'Ta-myt', dedicated to Prince Thutmose, found in the capital city of Memphis. Bronze figurines of Bastet with a cat head and human body are common in museums worldwide, and some Roman Period bronze images of cats show them suckling kittens.

S.Ikram

Cat bent over some food; Tomb of Nakht (TT52), Luxor

A. Oliveira

Cat eating a treat. The unlikely pose evokes a human sniffing a lotus; Tomb of Ria (TT159)

Leopard ■ *Panthera pardus* HBL 130–190cm TL 60–100cm SH 70–80cm WT35–85kg

DESCRIPTION Large, powerful cat with a spotted coat. Males generally larger than females. Dark orange-buff to pale yellow above with white underparts. Spotted throughout with spots on body arranged in variably sized and spaced rosettes, with the inside of each rosette slightly darker than the background colour. Underparts and limbs have smaller, round spots. Head large and powerful and more finely spotted than body. Chin white. Ears rather small with black and white behind. Tail long, spotted as body on basal half, then with incomplete black and white bands. Vocal, including loud, rasping call and various growls and snarls. **DISTRIBUTION** In Egypt the natural distribution in historical times is unclear. By the last century, due to heavy persecution confined to the mountains of South Sinai, where it is almost certainly now locally extinct, although stone Bedouin Leopard traps can still be seen. One record from the Western Desert in 1913. May still exist in the Gebel Elba region in the south-easternmost part of the country. The Leopard is today still found over much of sub-Saharan Africa, across the Middle East and Arabia, and to South and South-east Asia north to Russia. **HABITAT AND HABITS** Ranges over a wide range of open habitats, but needs cover for concealment and hunting. In Sinai and elsewhere also in mountains. Solitary, with a home range that is defended. Takes a wide variety of prey up to the size of large antelopes, but generally smaller; historically in Sinai probably Nubian Ibex. Very elusive and the spots render it almost invisible in sun-dappled vegetation. **SIMILAR SPECIES** The Cheetah (p. 64) is also probably extinct in modern Egypt; it is lankier and slimmer, with solid black spots and blackish 'tear marks' running down the face from the eye.

CC Byrdyak

Leopard *Aby* or *Bameh*

Leopards appear in hunt scenes, in rows of wild animals, and quite often in scenes relating to tribute from more southern parts of Africa. They were feared as big cats and valued for their pelts. The names for Leopard and Cheetah in Egyptian were used interchangeably, presumably the idea being that the image would indicate which animal was meant. When the paint survives on material remains, the Leopard's rosette marking differentiates it from a Cheetah, but when the paint is missing it is harder to separate the two. Generally, Cheetahs are shown with their necks straight up, emphasizing their small heads, and Leopards are shown with their heads lowered, as is more common in nature. **PRACTICAL USES** It is possible that Leopards were kept as pets, but less commonly than Cheetahs, which are easier to tame. Leopard skins were highly prized in religious contexts. **MYTHOLOGY/ICONOGRAPHY** Leopard skins were part of the official garb of *sem*-priests, who are seen officiating particularly at funerals, although such skins are also seen being worn by other priests. Leopards symbolized the wilderness that needed to be tamed so that order could prevail. Their statuettes in tombs, covered in a black resinous substance, and associated with a figure of a king, evoked their nocturnal aspects, and the need to control or harness these so that the sun could be reborn and order maintained. **HIGHLIGHTS** In hunt scenes in tombs. Occasionally in Old Kingdom pyramid temples in causeways. The pelts being worn by priests are common, notably in New Kingdom tombs, particularly on the west bank of Luxor, with a fine representation in the tomb of Tutankhamun in the Valley of the Kings. A garment made of linen replicating Leopard skin with the rosettes translated into gold, with a gold head with glass inlays, was part of the funerary equipment of Tutankhamun. A statue showing King Amenemhat III as a priest in the Egyptian Museum also shows him wearing a Leopard skin. Gilded figurines of King Tutankhamun astride a Leopard that was coloured completely black with a resinous substance evoking the afterworld, were found in his tomb.

S. Ikram

Leopard being brought from Punt to Hatshepsut, Deir el-Bahari

S. Ikram

Leopard being brought by Nubians; Tomb of Rekhmire (TT 100)

Lion ■ *Panthera leo* HBL 1.5–2.3m TL 1m SH > 1.2m WT 150–225kg (F smaller)

CC J. Wimmerlin

DESCRIPTION Very large cat, tawny throughout with distinctive mane in adult male. Very large but distinctly feline and sexually dimorphic. Short haired and unpatterned except in infants. Tawny beige with paler underparts. Ears broad and rounded. Chin and throat pale to white. Male develops a mane, fully developed at 5–6 years and variably coloured. Tail long and uniform with dark terminal tuft. Cubs have shaggier and dappled coats. Voice includes the characteristic loud roar, but also growling and purring vocalizations. **DISTRIBUTION** In Egypt, the natural distribution in historical times is unclear. Seen in Predynastic portrayals from the southwestern Western Desert. Recorded as mummies from ancient Egypt. Extinct in modern Egypt. Elsewhere in declining numbers over much of sub-Saharan Africa, where everywhere threatened by habitat destruction, persecution and conflict for resources with pastoralists. Races in north-west Africa and the Cape now extinct, and now very local in West Africa. In Asia previous populations over much of the Near East and India, but now confined to a single population in the Gir Forest, India. **HABITAT AND HABITS** Ranges over semi-desert to savannah to open forest. Absent from true desert (decline in Egypt probably due to retreat of the savannah as climate warmed). The most social of the cats, generally living in prides, with a dominant male/males and females, subadults and young. Territorial. Usually active in early morning and evening, but also nocturnal. Preys largely on medium to large mammals, especially ungulates, but other prey recorded, including scavenging. Females take the most active role in hunting. Sight and hearing acute. Smell good. **SIMILAR SPECIES** None. It is unclear when the Lion disappeared from modern Egypt. The cultural impact of the Lion is felt not only in Egypt but across the areas where it has long been extirpated, such as southeastern Europe and the Near East.

Lion *Maai* or *Ru*

Lions were not only roaming wild in Egypt, but also by about 2500 BC were kept in hunting enclosures and menageries, as well as being royal pets. Lions were symbols of rulership from the dawn of Egyptian civilization, trampling and devouring enemies. At Abydos, the burial place of Egypt's first rulers, Lions were interred near kings. Both Tutankhamun and Ramesses II are shown with Lions as pets, accompanying them on hunts and in battle. Both Lions and lionesses figure prominently in Egyptian art, playing a part in the sacred as well as secular lives of the ancient Egyptians, appearing in paintings, reliefs, on furniture, and as elements in jewellery. Sometimes when lionesses are depicted as goddesses, they are shown with a stylized mane. From the Ptolemaic Period at the site of the Bubasteion at Saqqara, dedicated to the cat goddess Bastet, the mummies of a Lion and at least one cub have been found. **PRACTICAL USES** Lions were hunted and were prestige pets for kings. A Lion pelt was found in the tomb of Tutankhamun, probably serving as a carpet in one of his six chariots. **MYTHOLOGY/ICONOGRAPHY** Lions play an important role in Egyptian iconography, being a manifestation of the king, but also of several deities. Sphinxes, partly Lions and partly humans, were protective symbols associated with solar deities, notably the sun god Re. They served as guardians of temples and cemeteries. The god Ruty, defining the horizon, was shown as two back-to-back Lions, and worshipped at Leontopolis (Tell el-Moqdam) as well as other sites with solar connections. Other gods with Lion imagery are Apademak, Mahes, Tutu (like a sphinx) and Aker. Goddesses associated with lionesses are Sekhmet, associated with protection and warfare, Mekhit, Tefnut, Mut, the queen of the gods and consort of Amun, and Wadjet, a key protector of the King. Bastet in her calmer form was a cat, but when enraged she was a lioness. **HIGHLIGHTS** The great Sphinx at Giza and the avenues of sphinxes at other temples. The stela between the Sphinx's paws records a Lion hunt as well as a divine prophecy that made Thutmosis IV king. Ruty is frequently pictured in the tombs of the Valley of the Kings and at the tombs of the workers in Deir el-Medina. Statues of Sekhmet and Mut are found in many museums, as well as the temples of Mut at Karnak and that of Amenhotep III (Kom el-Hitan) on the west bank of Luxor. Leonine heads adorn Tutankhamun's throne and other furniture, and he is shown hunting Lions on one of his painted wooden chests. In depictions of Ramesses II's Battle of Qadesh at Abu Simbel and the Ramesseum in Luxor, the Lions are shown running beside the royal chariot or being stabled close to the horses. In Saqqara Lions are shown being hunted in hunt scenes.

J. M. Galán

A striding Lion at Karnak temple

Cheetah ■ *Acinonyx jubatus* HBL 110–160cm TL 65–80cm SH 80cm WT 40–60kg

A. Main

DESCRIPTION Lanky, slimline big cat with a spotted coat. Large, slender and small headed, with long, slender limbs and tail. Above, flanks and outsides of legs buffish to yellow-buff, spotted throughout with solid, dark, round spots never arranged in rosettes as in the Leopard (p. 60). Underparts paler with diffuse spotting. Fur short and dense but with an erectile mane along back. Head proportionately small with flat top and small, rounded ears. Densely spotted with a diagnostic 'tear mark' running from eye down side of face. Muzzle short. Tail long and full, broader at tip than base, and ringed black and white towards tip. Tip white. Unlike most cats claws only semi retractile. Cubs with a pale back possibly mimicking the Honey Badger *Mellivora capensis*. **DISTRIBUTION** In Egypt the natural distribution in historical times is unclear. By the last century, perhaps due to heavy persecution of its natural prey, primarily the Dorcas Gazelle, it was confined to the Qattara Depression in the northern Western Desert, where it clung on until the last unconfirmed records in 1994. Records from the 1940s in North Sinai. Elsewhere, Mauritania, throughout sub-Saharan Africa in suitable habitat. Absent from much of West and Central Africa, and with an isolated population in Iran. Previously more widespread in the Middle East and South Asia. **HABITAT AND HABITS** Open plains and savannah to semi-desert. In the Qattara Depression largely around *Acacia* groves. Hunts by running down prey such as gazelles, small to large mammals and birds, after an explosive sprint of up to 100km/ph, after which the prey is toppled. Hunting is by sight. **SIMILAR SPECIES** The Leopard is much stockier and shorter limbed, with completely different habitat requirements. Its spots are arranged in rosettes, unlike the solid spots of the Cheetah.

Cheetah 𓇋𓃀𓇌𓃮 *Ahby*

The names for Cheetah and Leopard are generally used interchangeably, with the image indicating to the viewer which animal was meant – and sometimes perhaps either or both were intended. This is further confused when images show the rosette markings of a Leopard on the body, but the face is painted or carved with the distinctive markings found on Cheetahs. While Leopards probably were more common in the landscape of ancient Egypt, Cheetahs were known, too, and appear in hunt scenes. In the New Kingdom they are shown being brought as tribute from further south in Africa. **PRACTICAL USES** The animals would have been hunted and could have been prestige pets. Indeed, today in parts of the Middle East the very wealthy import them as pets as they are the easiest of the large cats to tame. The pelts could have been used as clothing in religious rituals as well as luxury coverings for furniture or floors. **MYTHOLOGY/ICONOGRAPHY** Cheetahs do not have a significant role in Egyptian religion, although they, with other big cats, could be associated with feline deities. In hunt scenes, their capture was a symbol of control over chaos. **HIGHLIGHTS** It is sometimes difficult to distinguish between Egyptian representations of Cheetahs and other big cats when the painted definition is missing. However, they seem to be featured in the tomb of Rekhmire (TT100) and in Hatshepsut's temple at Deir el-Bahari, where both Leopards and Cheetahs are brought, with the latter being distinct from the former due to their longer necks and the proud, erect stance of their heads.

F. Dzikowski

Cheetah being brought to Thutmose III; Tomb of Rekhmire (TT100)

African Wild Ass ■ *Equus africanus* HBL 195–305cm TL 40–45cm SH 115–135cm WT c. 275kg

Race occurring (that occurred?) in Egypt is the Nubian Wild Ass *E. a. africanus*

DESCRIPTION Small equid and typically donkey-like in form. Coat short and smooth. Upperparts and flanks pale grey tinged ruddy. White below. Dark stripe along back of spine to base of tail and with a transverse stripe across the shoulder generally present. Legs paler and without striping. Head large with pale muzzle and jaw and white ring around eye. Ears large, pale inside, grey on back. Mane running from between ears to base of neck, with short, erect hairs. Tail long with dark tail-tuft. Voice as in the braying of the domestic donkey. This is very loud and can carry for up to 3km. **DISTRIBUTION** In Egypt, historically through the southern Eastern Desert, but now probably extinct with only recent records from Gebel Elba in the far south-east of the country. Historically also found through northeastern Sudan south to Eritrea, but now almost certainly extinct. The factors behind extinction include hunting, competition for very limited desert resources with increasing herds of domestic livestock, and hybridizing with Domestic Donkeys. **HABITAT AND HABITS** Desert and semi-desert regions including mountainous areas. Adept climbers over rocky terrain. Can go without water for up to three days. Habits unstudied in Egypt, but elsewhere males may be solitary or form bachelor herds. Females also in herds, sometimes with young. Herd size generally 6–12 individuals, but up to 50 recorded in the past. Food generally grass and other grazed vegetation, rarely leaves. Mature males defend a territory. Gestation 330–365 days. Generally one foal. **SIMILAR SPECIES** The Nubian Wild Ass is the only wild equid recorded in Egypt. The Somali Wild Ass *E. a. somaliensis* ranges further south through Somalia, Ethiopia and Eritrea but is now Critically Endangered. It differs in having a series of horizontal stripes on the legs. This is the wild ancestor of the domestic donkey, with domestication thought to have occurred around 6,000 years ago. Today, seemingly feral donkeys may be found in areas along Egypt's Eastern Desert but are invariably domestic animals, for example in the Wadi Gimal Protected Area.

CC M. Gikas

Nubian Ass *E-a-ah*

The wild ass, with its onomatopoeic name, features in desert scenes, and Domestic Donkeys were commonly featured in Egyptian tombs of all eras. **PRACTICAL USES** As a hunted animal for food and sport; Domestic Donkeys were used for transportation and as beasts of burden. **MYTHOLOGY/ICONOGRAPHY** The wild ass was associated with the chaotic forces of the desert that needed to be tamed, and also with the god Seth. Domestic Donkeys do not seem to have had a particularly negative connotation, although the association to Seth could be extended to them. **HIGHLIGHTS** On Tutankhamun's painted box (Carter No. 021, JE 61467) the king hunts wild creatures, including the Nubian Ass. Other images of hunting these animals are found in Theban tombs. Domestic donkeys appear in agricultural scenes or tribute scenes in tombs of the Old, Middle and New Kingdoms.

S. Ikram

Above: A group of donkeys going to work in the fields; Tomb of Paheri, el-Kab

Right: A donkey carrying a basket of young gazelles; Tomb of Two Brothers, Saqqara

S. Ikram

Wild Boar ■ Sus scrofa HBL 100–146cm (F) 105–167cm (M) TL 16–30cm SH 59–109cm WT 30–80kg (F) 100–130kg (M)

CC Haberdoedas

DESCRIPTION Large, stocky, pig-like mammal with proportionately large head and short legs. Males generally significantly larger than females. Body with coarse, bristly fur dark blackish-brown to greyish-yellow in adults with a dorsal crest. Legs short with pointed hooves. Head large but narrow with long muzzle; naked snout. In male upwardly and outwardly pointed tusks of upper canines. Uniquely, male Wild Boar have heavy shields of protective skin on the shoulders and chest, which deflect blows from rivals' tusks. Ears large, erect and pointed. Tail short. Young very different from adults, with horizontal stripes of brown and yellow. Snuffles and grunts when foraging. Barks when alarmed. **DISTRIBUTION** Extinct in modern Egypt. In prehistoric times probably distributed along the entire Delta and Valley. Hunted extensively in ancient times, and the Wild Boar in Egypt became extinct in the wild c. 1900, with the last stronghold probably being Wadi Natroun. A single animal survived in Giza Zoo until 1912. Elsewhere found across Europe through the Middle East to central, south and southeastern Asia. Also North Africa. **HABITAT AND HABITS** Mostly deciduous woodland and scrub, but in Egypt seems to have generally been recorded from reedbeds and swampland. Largely nocturnal and crepuscular. Males generally solitary except during the rut. Females found in herds with young. Litter size can be up to 10, and breeding is every year. Omnivorous, feeding on all types of vegetable matter, but also carrion, eggs, beetles, earthworms and even small mammals. **SIMILAR SPECIES** None in Egypt, where even the Domestic Pig *S. domesticus* (considered by some to be just a race of the Wild Boar) is rarely seen, especially after outbreaks of Swine Flu. The species' extinction in Egypt is probably a combination of hunting and habitat destruction. The pig was domesticated by the ancient Egyptians since Neolithic and into Dynastic times.

Wild Boar [hieroglyphs] *Reri* or [hieroglyphs] *Shai* (also used for pig)

Rarely featured in Egyptian art, although boar and pigs are known from the archaeozoological record. **PRACTICAL USES** Boar were hunted, provided meat and fat, and the tusks were used in weaponry, as inlays, bodily decoration and possibly in tool production. **MYTHOLOGY/ICONOGRAPHY** Boars and pigs were associated with Seth, the god of chaos, and frequently had negative connotations. Thus, images of boars being speared or harpooned are indicative of the victory of the god Horus over Seth and the maintenance of cosmic order (*maat*). In the Book of the Gates, a funerary text, images of a monkey hitting a boar are interpreted as being the triumph of the sun over darkness. Curiously, many figurines of pigs have been found dating to the Predynastic Period, and their role remains unclear. It is possible that they relate to an epithet of the sky goddess Nut: The Sow Who Eats Her Piglets or The Great Sow, referring to night time, when the sun travelled through the body of Nut to be reborn. Classical writers such as Herodotus, writing in the fifth century BC, suggest that pork was forbidden to priests, which might have been true only for priests of certain gods, or at particular occasions, as pig bones have been found in temple sites, as well as in middens of both commoners and elites. **HIGHLIGHTS** Several figurines of boar/pigs in faience, clay and other materials, of Predynastic date, are on display in the Egyptian Museum. Pigs emerging from the marsh are featured in the tomb of Khety (Tomb 17) at Beni Hasan, and trampling grain (tombs of Renni, Tomb 7, and Paheri, Tomb 3, at el-Kab,). In Edfu temple, images show the god Horus spearing the god Seth in his guise as a boar/pig.

S. Ikram

Wild Boar in the prow of a boat; Tomb of Ramesses II (KV 7), Luxor

S. Ikram

A group of pigs, descended from Wild Boar; Tomb of Paheri, el-Kab

Hippopotamus ■ *Hippopotamus amphibius* HBL 280–370cm (F) 320–420cm (M) TL 35–50cm SH 130–165cm WT 510–2,500kg (F) 650–3,200kg (M)

DESCRIPTION Huge, semi-aquatic, pig-like mammal with proportionately large head and short, thick legs. Males generally significantly larger than females. Body virtually hairless with bristles only on muzzle and tail. Overall purple-grey to pinkish-black, and paler pinkish below. Hippos exude a reddish fluid from extensive skin glands, thought to serve as a 'sun cream'. Legs short with splayed toes. Thick necked with large head and huge snout. Eyes, ears and nostrils all on top of head, adapted for life in the water. Males especially have huge canines and incisor teeth. Ears small and rounded. Tail very short and stubby. Very vocal, especially in large agglomerations, with a variety of honks, wheezes and roars. **DISTRIBUTION** Extinct in modern Egypt. Historically found in Upper Egypt from just north of Aswan and on the Damietta branch of the Nile in the Delta. Probably disappeared from the Delta sometime in the early seventeenth century and from Upper Egypt perhaps in the early nineteenth century. Elsewhere restricted to Africa south of the Sahara, where it is found along many major river systems, but is largely absent from the south and south-east. **HABITAT AND HABITS** Requires permanent fresh water and extensive grazing. Spends the day wallowing, then emerges to graze at night on a variety of grass species. Large conglomerations can occur, especially in the dry season. Adult males defend an area of river bank and their females against rival males, but tolerate subadults. Dominance is asserted by yawning displays, where the lower jaws are clashed together. Subservience is shown by prostration, urination and so on. **SIMILAR SPECIES** None in Egypt. The only other hippo species is the Pygmy Hippopotamus *Hexaprotodon liberiensis*, which is confined to the tropical forests of West Africa.

CC H. Cornelissen

Hippopotamus *Nehes* or *Deb* or *Khab*

S. Ikram

Tomb of Idut, Saqqara

Hippopotamuses are portrayed from early in Egyptian history, particularly in tombs as well as in temples, in addition to being a popular item of sculpture in the Middle Kingdom. **PRACTICAL USES** In Predynastic times hippopotamuses were hunted by groups of humans using harpoons and spears. Their meat could be eaten and the fat was rendered and used for cooking as well as for making ointments (medical, cosmetic and practical). At the site of Maadi (4000–3500 BC), hippo leg bones were used, in conjunction with mud, to make stairs. Their ivory was used to make or decorate boxes, furniture, small vessels such as kohl pots, and clappers used for sacred and secular music. The thick hide might have been used to make objects such as shields, but these have not survived. **MYTHOLOGY** Male hippos were associated with the god Seth, god of storms, deserts, disturbances in nature and chaos. Thus, killing the males not only had a practical function of culling overly large herds and producing meat and by-products, but also symbolically showed the triumph of order/control (*maat*) over chaos (*isfet*). Images of groups of men hunting hippos, as well as kings harpooning these animals, are commonly found in Old Kingdom (and later) tombs and memorial temples. In the Ptolemaic temple at Edfu, Seth is shown as a hippo being speared by the god Horus. Female hippos were associated with the goddesses Taweret, Ipet and Reret. All three were benign and protective of women and children. Taweret in particular was revered as a protector of children, mothers and women in childbirth, with chapels being dedicated to her. Many amulets featuring this goddess are found in Egypt from both settlements and cemeteries. In addition, blue faience hippo figurines, painted with images of lotus and papyrus, were found in tombs of the Middle Kingdom. In order to render them harmless, their feet were often broken off, although kept next to the figurines. **HIGHLIGHTS** Hippo-hunting scenes are particularly popular in the Old Kingdom and can be seen in the mastaba tombs of Idut and Mereruka at Saqqara. Blue faience hippos were popular in the Middle Kingdom and examples of these can be found in the Egyptian Museum, the Grand Egyptian Museum and the Metropolitan Museum of Art, where their hippo has been nicknamed 'William', and is a mascot for the museum.

Dromedary ■ *Camelus dromedarius* HBL 2.0–3.3m TL 50–70cm SH 1.7–2.2m WT 300–450kg (F) 400–700+kg (M) (Arabian Camel, One-humped Camel)

DESCRIPTION Large, long-legged, long-necked ungulate superbly adapted to life in the desert. Males generally larger than females. Tall with laterally compressed body, narrow chest and one hump of fatty and fibrous tissue. Coat generally uniform brown to sandy-brown, but can be almost white, much darker or rarely piebald. Fur short but longer on throat, shoulders and hump. Callosities of thick skin on joints and chest. Neck long and curved. Head has a long, downcurved muzzle with split upper lip, large eyes with double eyelashes and bushy eyebrows. Ears small and rounded. Legs long, accentuated by lack of tensor skin between thigh and body, and with two-toed broad, padded hoofs. Tail short. **DISTRIBUTION** Extinct in the wild for perhaps 2,000 years but with feral populations especially in Australia. Originally found in western Asia to the Arabian Peninsula, and domesticated either there or in Somalia. Today, the Dromedary is found as a domestic animal over much of northern Africa south to the Horn and to East Africa, as well as across Arabia and to western Asia. In Egypt widespread as a domestic animal. **HABITAT AND HABITS** Herds, generally of females led by a mature male, or of bachelor males, can be found roaming in desert and semi-desert environs and also as fully domestic animals in fertile agricultural areas. Free-ranging animals are highly mobile, travelling up to 50km in a day and browsing on a variety of plants, including thorny trees and shrubs. The hump is a fat reserve and camels can go for up to 15 days without water. The double eyelashes, bushy brows and so on, protect the animal when feeding and in sandstorms. **SIMILAR SPECIES** While the Dromedary is popularly associated with Egypt, it arrived in Egypt relatively late in pharaonic times.

Shutterstoc/O. Prosicky

Dromedary No word for dromedary in ancient Egyptian

No one is quite sure when Dromedaries came to Egypt, but it has been suggested that they arrived at some time in the eighth century at the earliest, but more likely the seventh/sixth centuries BC, becoming popular subsequently. Once they had arrived, camels became a mainstay of desert travel due to their ability to go without food or water for extensive lengths of time, carrying capacity and the fact that their hooves are well suited to traversing soft sand. Outside of rock art, few images of camels are found in Egyptian art, although a notable exception is a faience vessel in the shape of a camel, now in the Egyptian Museum. **PRACTICAL USES** Transportation of people and goods, racing animals, and a source of food and bones that could be used for tools and jewellery. **MYTHOLOGY/ICONOGRAPHY** None. **HIGHLIGHTS** Rock art in Aswan, the Eastern and Western Deserts, and the oases feature camels. Camel graffiti appears on temples, notably those in Nubia that are visited on the Lake Nasser Cruise.

S. Ikram

Camels in rock art, Wadi Mineiha, Eastern Desert

S. Ikram

Camels in rock art, Wadi Mineieha, Eastern Desert

S. Ikram

Camels in rock art, Wadi Mineieha, Eastern Desert

Giraffe ■ *Giraffa camelopardalis* HBL 3.5–4.8m TL 76–110cm SH 270–330cm TH 3.5- 5.2m WT 450–1,180kg (F) 1,800–1,930kg (M)

DESCRIPTION Unmistakable. Huge, very tall (to more than 5.2m in male) herbivore with very long neck and legs. Males significantly larger than females. Body very variably patterned between the various races and between individuals. Generally, ground colour pale beige spotted and blotched throughout in a 'crazy-paving' pattern. Some races, such as the Reticulated Giraffe G. *c. reticulata* of northern Kenya, have large, uniform blotches separated by very narrow lines, while other populations have less close-fitting blotches and are much paler in colour. Legs long with broad and rounded hooves. Neck very long with prominent mane. Head has narrow, tapered muzzle with long, blue-grey tongue. 'Horns' present in both sexes. Eyes large and dark. Tail long with a prominent tip. **DISTRIBUTION** Extinct in modern Egypt, probably disappearing as the climate warmed 8,000–10,000 years ago. Today, still found over much of east, central and southern Africa, with more isolated populations in the west and south, but increasingly confined to reserves and Protected Areas. **HABITAT AND HABITS** An animal of savannahs and open woodland with ample deciduous trees, especially *Acacia* species, for food. The Giraffe is a browser. Diurnal, feeding in early morning and later afternoon, and retiring between these times. Vulnerable at waterholes, where it has to spread the front legs in order to lower the head to the water. Does not live in stable herds, with the mother–calf relationship the only lasting social bond. Young males are involved in 'necking' duels that may establish a hierarchy. Despite their size Giraffes are adept runners, reaching 60km/ph, using a distinctive loping gait. **SIMILAR SPECIES** The Giraffe is unmistakable. Its only close relative is the much smaller and darker Okapi *Okapia johnstoni*, limited to the Congo rainforest.

Bernard & Tara O'Kane

Giraffe *Ser* or *Memy*

Giraffes became extinct in Egypt by about 3000 BC and were only present thereafter in menageries, as attested by zooarchaeological finds dating to the Ptolemaic Period. **PRACTICAL USES** Giraffe tails could be used as fly whisks, the skin could be used in furniture, the meat could be consumed and the bones used to make tools or weapons. **MYTHOLOGY/ICONOGRAPHY** The Giraffe does not play a formal role in Egyptian mythology, although some scholars have suggested that it was a solar creature. The word for Giraffe, *ser*, also means to foretell, because due to its great height, the giraffe can see far away and knows what is to come. Giraffes feature prominently in the rock art of the Eastern and Western Deserts, especially the latter. They might also have been associated with being able to foretell rain in Predynastic times. **HIGHLIGHTS** Nubians are shown bringing a Graffe as a gift/tribute to King Thutmose III in the Theban tomb of Rekhmire (TT 100). Giraffes feature in rock art, sometimes in hunt scenes.

A. Oliveira

Tomb of Rekhmire (TT 100)

Scimitar-horned Oryx ■ *Oryx dammah* HBL 195–220cm TL 45–65cm SH 110–135cm WT 135–180kg

DESCRIPTION Large, pale antelope with very long, back-curved horns. Adult very pale whitish throughout with no or very indistinct flank-stripe. Pale orange to chestnut on neck and shoulders, and along back to tail-tuft. Legs white with orange patch on carpus. Head whitish with pale orange to chestnut along top of snout, between horns, and patch running vertically through eye. Ears quite large and pale whitish. Tail to below hock, whitish, orange-brown above to terminal tuft. Horns present in both sexes, very long and curved back. Ringed. Thicker and heavier in male, generally thinner and longer in female. Not vocal but has a low grunt and a bleat if threatened. **DISTRIBUTION** Extinct in modern Egypt. Historically with records from the Western Desert from the nineteenth century from around Siwa, Kharga, Dakhla, Wadi Natrun and Fayoum. Last sight record from south of Mersa Matruh in 1972. Formerly across much of North Africa, from southern Morocco, south to Senegal, east across Mali, Niger, Chad and Libya, to Sudan and Egypt, possibly to Sinai. Decimated by climate change, hunting and competition with domestic livestock. Declared extinct in the wild in 2000, but with captive breeding programmes in a number of countries, including Tunisia, Morocco, Senegal and Chad. **HABITAT AND HABITS** Species of arid desert and semi-desert. Drinks when water is available but can survive long periods without doing so by gleaning enough moisture from its diet of grass, shoots, leaves and fruits, and from dew. Sociable, living in herds of up to 20–30 or more in the past. Active mainly in early morning and evening, lying up by day in shade if available. **SIMILAR SPECIES** The Arabian Oryx *O. leucoryx*, also known as the White Oryx, is a related species and one that has suffered a similar fate to the Scimitar-horned Oryx. It is slightly smaller, white marked with dark brown on the head and neck, and with brown legs. The horns are long, ringed, and straight or only slightly curved. Unknown in modern Egypt, it was recorded from Sinai as recently as the early 1800s. Due to similar factors to those applying to the Scimitar-horned Oryx, it retreated to the vastness of the Arabian Desert, where the last four wild animals were taken into captivity in 1972. It has since been reintroduced into a number of Gulf countries, Syria and Jordan.

A. Main

Oryx 𓌳𓄿𓌉𓃵 *Ma-Hedj*

S. Ikram

Tomb of Idut, Saqqara

S. Ikram

Tomb of Mereruka, Saqqara

Wild animals that are featured in tomb and temple scenes, but not in three-dimensional art. **PRACTICAL USES** Oryxes were eaten, hunted for sport, portions could be used in medicine and magic, and the horns could be used to make tools, weapons and musical instruments. **MYTHOLOGY/ ICONOGRAPHY** Oryx, like many other antelope, were identified with wild creatures that needed to be tamed and controlled so that order (*maat*) could be maintained. A specific ritual associated with royal control over chaos (*isfet*) seemed to involve the beheading of an oryx. Additionally, in the Ptolemaic era in particular, the powers of chaos were sometimes personified by an oryx. **HIGHLIGHTS** First appearing in Predynastic rock art, oryx featured in hunt, force-feeding and offering scenes from the Old Kingdom onwards (there are several examples in the Memphite mastabas). Examples of the king controlling the wild by beheading an oryx can be seen in Amenhotep III's birth room in Luxor Temple, as well as in temples, such as that of Edfu.

Addax ■ *Addax nasomaculatus* HBL 120–175cm TL 25–35cm SH 95–115cm WT 60–135kg

CC G. Todd

DESCRIPTION Rather small, very pale antelope with long horns in an open spiral. Size of a donkey but much more graceful. Sexes similar, with male slightly larger than female. Largely pale grey-brown, white to whitish on underparts, rump and legs. Paler in winter than summer and with age. Brown on crown in front of horns. Rather short muzzle, grey-brown with pale band below eyes. Ears white behind. Tail longish, extending to hock, whitish and with white terminal tuft. Horns present in both sexes, long and slender with an open spiral and up to 89cm long in mature males. Ringed for the lower two-thirds. Horns of younger individuals straight or weakly spiralled. Thicker and heavier in male, generally thinner and longer in female. Not vocal but grunts recorded. **DISTRIBUTION** Extinct in modern Egypt. Historically with records from the Western Desert from the nineteenth century, east to Fayoum and the margins of the Western Oases. Last reports of a live Addax in Egypt from a sight record in 1900 west of Alexandria and 1931 west of what is now Lake Nasser. Perhaps in the Eastern Desert in pharaonic times. Formerly across much of North Africa, from Mauritania south to Senegal, east across Mali, Niger, Chad and Libya, to Sudan and Egypt. Decimated everywhere by hunting and competition with domestic livestock. Now with wild populations only in Niger and possibly Chad and Mauritania. Captive-breeding progammes in a number of countries, and reintroduced into Tunisia and Morocco. **HABITAT AND HABITS** The Addax is a species of arid desert, dunes and semi-desert. Does not require standing water, obtaining moisture from food and dew, but will drink when water is available. Feeds largely on grass and leaves. Nomadic and sociable, living in herds of up to 20, but reportedly many more in the past. Active mainly in early morning and evening, and also at night, lying up in the day in shade if available. **SIMILAR SPECIES** None in modern Egypt. Two other spiral-horned antelope species reach Sudan and Eritrea and were more widespread in historical times. The Greater Kudu *Tragelaphus strepsiceros* is a far larger species with more tightly spiralled horns and white vertical stripes on grey-brown flanks. The Lesser Kudu *T. imberbis* is smaller and slimmer. It is a dry-country species reaching South Sudan and Ethiopia, and may have occurred in Arabia to recent times.

Addax 𓃘 *Nwdjiw*

Addax are pictured in Egyptian tombs, particularly those of the Old Kingdom. The skeletal remains of these animals have also been found in various parts of Egypt and Nubia. **PRACTICAL USES** Addax provided food and sport, in terms of hunting. Their skin could be used for furniture, clothing or other objects, and their horns may have been used to make tools. **MYTHOLOGY/ ICONOGRAPHY** No deities are associated with the Addax. They are animals that are part of the desert landscape that need to be controlled and tamed. **HIGHLIGHTS** Drawings of Addax adorn ceramic vessels of the Predynastic Period that are found in many museums in Egypt and elsewhere. Addax are found in offering and hunt scenes in the Old Kingdom tombs at Saqqara, and appear in the rock art of both the Eastern and Western Deserts as well as Nubia.

S. Ikram

Tomb of Kagemni, Saqqara

Dorcas Gazelle ■ *Gazella dorcas* HBL 88–100cm TL 10–18cm SH 55–65cm WT 14–20kg

Races occurring in Egypt G. *d. dorcas* and G. *d. isabella.* **DESCRIPTION** Small, quite thickset gazelle with slender limbs and striking head pattern. Males slightly larger than females. Upperparts very pale to pale red-brown, darker in winter and in some subspecies. Darker flank-stripe separates from white underparts. Legs as upperparts, inner side white. Head distinctly marked with darker central strip from between horns to nasal region, bordered by white stripe running from below horns and around eye. Both sexes have horns. In mature male horns rather thick and lyrate, curving back and up and often curved in at tip. In female horns more slender and straighter. Tail dark on upperside contrasting with white rump. Not vocal, but may utter a short bark. **DISTRIBUTION** In Egypt historically over much of the Western Desert, Eastern Desert and Sinai. Much depleted everywhere, but still found in Protected Areas such as St Catherine's and Wadi Gimal. Elsewhere in declining numbers over much of North Africa, south to the northern Horn of Africa and in the Naqab and Arabah Deserts. **HABITAT AND HABITS** Stony desert and semi-desert, coastal plains, and in the Eastern Desert and Sinai, vegetated wadi floors. In pairs and small herds. Mainly active in early morning and late afternoon, but may be nocturnal where persecuted. Feeds on grasses and other plant species, including *Acacia* species. Very wary, with excellent senses of sight, smell and hearing. **SIMILAR SPECIES** The Slender-horned Gazelle G. *leptoceros* is slightly larger than the Dorcas Gazelle and much paler (the alternative name is White Gazelle). The horns are long (up to 41cm in male) and only slightly curved. This is a much rarer species in Egypt, restricted to areas of the Western Desert and is a gazelle of vegetated sandy desert and dunes. In Egypt it has been devastated over much of its North African range by active hunting and competition with domestic livestock. The Mountain Gazelle G. *gazella* has a wide range from northwestern Africa across to Syria and the Arabian Peninsula. In Egypt it is known from sight records from the 1930s in North Sinai.

Shutterstock/T. Genes

Dorcas Gazelle *Gehes*

Gazelles, not clearly identified to species, are found on the walls of Egyptian tombs and temples, and 'gazelle' was also used as a name or nickname for women. Mummified gazelles are known from the site of Kom Mereh, and there is a possible gazelle mummy that was the pet of Isetemkheb D. **PRACTICAL USES** Food, pets. **MYTHOLOGY/ ICONOGRAPHY** Gazelles are not only shown in hunt scenes to be tamed, but are also associated with the goddess Anuket. She was the goddess of Egypt's southern border, particularly the area around modern Aswan, and was guardian of the king as well as a huntress. Also, the Near Eastern god Reshep wore a diadem with gazelle horns or a gazelle head centerpiece, stressing his association with the desert. **HIGHLIGHTS** In Giza and Saqqara, gazelles are found in hunt and offering scenes of mastabas. In Luxor they are shown not only in these contexts but also as pets, close to the chair of their deceased owner. Two possible gazelle mummies are located in the Animal Mummy Room of the Egyptian Museum.

S. Ikram

Gazelle being led; Tomb of Idut, Saqqara

S. Ikram

Tomb of Pabasa (TT 279), Luxor

S. Ikram

Gazelle being brought as an offering slung over a man's shoulders; a hartebeest foal is in the arms of the man behind; Tomb of Two Brothers, Saqqara

Nubian Ibex ■ *Capra nubiana* HBL 140–170cm TL 15–25cm SH 65–90cm WT 50–80kg

T. Holmgren

DESCRIPTION Robust, goat-like animal with backswept, curved horns especially long and heavy in mature males, which are significantly larger than females. Grizzled light to darker brown above, extending to upper half of legs. Flanks as upperparts. Underparts and insides of legs white. Leg pattern striking in both sexes, with black patches above and below each knee and the knee (pastern) white. White above hooves. Particularly in males, black along dorsal crest, chest, chin and muzzle. Females less strongly patterned. Mature males and some older females have a black beard. Ears brownish, dark inside and with white margins. Tail short and with black margins. Horns present in both sexes, but much longer and heavier in mature males. Horns long, backswept and strongly curved, robust at base with up to 30 raised knobs on front edges smoother towards tip. Recorded up to 138cm along curve. Emits a 'whistling snort' when threatened. Kids bleat. **DISTRIBUTION** In Egypt, formerly widespread over much of Sinai and the Eastern Desert south to Gebel Elba. Decimated last century by hunting but can still be found in South Sinai in the St Catherine's, Abu Galum and Taba Protected Areas and in the Eastern Desert in the Gebel Elba Protectorate and the Wadi Gemal Protected Area. Elsewhere, historically throughout the Middle East, including Saudi Arabia and Yemen and south to Sudan and Eritrea, but everywhere much reduced and fragmented. **HABITAT AND HABITS** Confined to rocky, mountainous terrain, including cliffs, canyons and wadis. Active mainly by day in early morning and evening. Sociable, in herds of up to 40 animals, but generally in smaller groups. Adept climber. Needs standing water. Males clash aggressively for females, with dominant male mating with his harem. **SIMILAR SPECIES** Formerly considered as a subspecies of the Alpine Ibex C. *ibex*. The Domestic Goat is much smaller, very variably coloured and with different horns. The Barbary Sheep (Aoudad) *Ammotragus lervia* is a wild sheep similar in size to the ibex but rather uniform orange-brown, paler below and with a mane along the underside of the neck. The horns are heavy, curved, almost circular in mature males, but lack the knobs of the ibex. In Egypt now probably confined to the southern Eastern Desert and the remotest areas of the southwestern Western Desert.

Nubian Ibex *Niaw*

Ibex feature prominently in two-dimensional art starting in the prehistoric period, and continue to be featured throughout Egyptian history, particularly in hunting scenes. **PRACTICAL USES** Food, medicine and magic. Their bones, hide and horns could also be used. The horns, with the addition of a wooden sound-box, might have been used to make a form of harp. **MYTHOLOGY/ICONOGRAPHY** Ibex were part of the group of wild desert animals that needed to be controlled so that order and balance (*maat*) reigned. The ibex's horn, due to its pronounced corrugations, was associated with the hieroglyph for 'year', originally a palm-rib with ribs, and thus the ibex was linked to New Year festivals. **HIGHLIGHTS** The rock art of the Eastern Desert is rich in ibex imagery, and both Eastern and Western Desert rock art also features Barbary Sheep. The latter are shown in hunt and offering scenes in tombs, but less frequently than ibex. The mastaba tombs of Giza and Saqqara, in particular, contain images of ibex being hunted and also being brought as offerings and force-fed for consumption. Hunting scenes featuring these animals continue into the end of Eighteenth Dynasty. They are also pictured in temples, being slaughtered and offered to the gods. Sculpted ibex adorn one of the Egyptian alabaster vessels belonging to Tutankhamun. Healing and protective statues (*cippi*) show the god Horus as a child standing on a crocodile and holding snakes, scorpions, ibex and other antelopes in his hands, with texts carved over the statue that protect against harm caused by wild and venomous creatures. Water, milk or wine would be poured over the statue, absorb its power, then drunk, so that the power of Horus and his protection against these wild creatures could be absorbed.

S. Ikram

Tomb of Two Brothers, Saqqara

Cape Hare ■ *Lepus capensis* HBL 300–465mm TL 49–102mm WT To 7kg
(Brown Hare)

CC Haplochromis

DESCRIPTION Large, superficially rodent-like mammal with very long, narrow ears, very large hindfeet and short tail, or scut. Upperparts ochre-brown to buff-grey with paler flanks and white underparts. Generally greyer in Sinai. Coat soft and dense. Legs and feet as body. Insides of legs whitish. Hindfeet very large – up to 100mm in length and sole with long hairs. Head has rather short, blunt snout and pale brown-buff throat. Eyes large and surrounded by pale eye-ring. Ears very large and elongated. Interior pale, otherwise buffish, sometimes with a dark tip. Ears held erect when alert and sitting upright, along the body when at rest in form. **DISTRIBUTION** In Egypt widespread and with up to five subspecies described. Found across the North Coast from Salum east and also in Siwa. Recorded around the Delta margins and in the Nile Valley in Upper Egypt. Status in Eastern Desert unsure but recorded from Wadi Degla and the Qena–Safaga road. Also, along the Red Sea coast and North and South Sinai. Elsewhere has a wide range over much of Europe, east across Arabia and the Middle East, and through India to China. Also found over much of sub-Saharan Africa. Introduced to the USA, Australia, New Zealand and elsewhere. **HABITAT AND HABITS** Open areas including desert and semi-desert where there is sufficient vegetation. Open farmland and margins, barley fields, wadi floors and rocky habitat and mountains in South Sinai. Largely nocturnal, but also active by day. Does not burrow, but lies up in a shallow depression known as a form. Very cryptic – best seen when erupting from cover jinking and zigzagging. Diet vegetarian. Hearing excellent. Sight acute and positioning of eyes gives it a very wide field of vision. Normally solitary. Deposits droppings in regular latrines. **SIMILAR SPECIES** While the Cape Hare looks like a large rodent, it is actually a member of the order Lagomorpha along with the other hares, rabbits and pikas. A key difference between rodents and lagomorphs is that lagomorphs have four incisors in the upper jaw and rodents just two. The Cape Hare is the only wild representative in Egypt. Similar in form are the unrelated hyraxes. The Rock Hyrax *Procavia capensis* and Yellow-spotted Hyrax *Heterohyrax brucei* are both found in Egypt in South Sinai and the Eastern Desert. These are colonial cliff dwellers feeding largely on *Acacia* leaves and seeds. Similar in size to the Cape Hare, they are readily distinguished by their short, rounded ears and lack of external tail.

Cape Hare 𓃹 *Wn*

Hares appear commonly in desert hunt scenes, as well as in scenes showing the yield of the hunt. They are most common in Old, Middle and New Kingdom hunt scenes. In hieroglyphs the hare had the sound 'wn', and thus appears in a multitude of words, including the name of King Unas (*wnis*) of the Fifth Dynasty, as seen in his pyramid at Saqqara. **PRACTICAL USES** Hares could be consumed and it is possible that their fur was used in clothing, although there is no evidence of its survival to date. **MYTHOLOGY/ ICONOGRAPHY** Hares do not have a prominent role in Egyptian religion, save in the form of amulets commonly found in the Middle Kingdom. Presumably these were indicative of the word *wnn* – to live, exist or be, written using the hare glyph. Hares do, however, appear in hunt scenes as wild animals that leap about and need to be controlled. They are sometimes shown being brought by hunters either in cages, as seen in the tomb of Khnumhotep II at Beni Hasan, or held by the ears by hunters, as pictured in the tomb of Nebamun in Thebes (Tomb 90). The hare was also a symbol of Egypt's fifteenth nome (province), now the area of Minya (Hermopolis Magna). **HIGHLIGHTS** In hunt scenes in tombs through the Eighteenth Dynasty. Occasionally in Old Kingdom pyramid temples in causeways. Figurines and amulets in museums.

S. Ikram

Hare hieroglyph; Tomb of Khuwy, South Saqqara

S. Ikram

Hare; Tomb of Nebamun, British Museum EA 37980

House Mouse ■ *Mus musculus* HBL 55–103mm TL 53–97mm WT 9.4–20.9g

DESCRIPTION Archetypal small brown rodent. Small mouse and the only Egyptian mouse with no spines. Very variable in size and colour. Generally grey through greyish-brown to brown with paler flanks and whitish underside, sometimes with buff tinge or even dark. Shade possibly linked to habitat. Head as upperparts; indistinct pale patch below eye and behind ear (more strongly marked in the spiny mice *Acomys* species. Ears relatively large and rounded. Tail as upperparts above and pale below; roughly equal to head and body length; untufted and with up to 200 tail-rings visible through fur. **DISTRIBUTION** Historically probably originated in Central Asia, but spread by humans globally. Introduced in Egypt, but long established. Found throughout, except desert regions, from the North Coast with records from North Sinai south through the Delta and Valley to Aswan and Abu Simbel. Recorded from all the Western Desert oases and from Wadi Natroun and Fayoum. Largely absent from the Eastern Desert but recorded along the Red Sea coast south to Marsa Alam. Isolated records from Sinai. **HABITAT AND HABITS** Commensal in cities, towns and villages; also farmland, groves, seashore and coastal desert, canal banks and irrigation ditches. Not in true desert, where replaced by the gerbils *Gerbillus* species and jirds *Meriones* species. Largely nocturnal but also active by day. Inhabits a shallow burrow with nest lining. Diet vegetarian and can be a pest in grain stores. Gestation 22 days with litter size of up to 10. Breeds throughout the year. **SIMILAR SPECIES** Two other mouse species are found in Egypt, both native. The Eastern Spiny Mouse *Acomys cahirinus* is larger than the House Mouse, with proportionally larger ears and a white patch below the eye and behind the ear. There are spines running from behind the shoulder to the tail-base. It is found throughout the Delta, Valley and along the Red Sea coasts and South Sinai. The Golden Spiny Mouse *A. russatus* is similar, but with upperparts orange to reddish-brown and with more extensive spines. It is restricted to rocky habitat in the northern Eastern Desert and South Sinai. It can best be seen on the trail up Mount Sinai and is active by day.

A. Main

House Mouse 𓊪𓈖𓅱𓅱 *Penu*

Mice (the House Mouse as well as other types of mice) were common animals in Egypt and considered serious threats to the fruits of the harvest. **PRACTICAL USES** Mice feature in medical texts, sometimes as cures for baldness or issues with the throat. In extremis they could serve as food. Different kinds of mice have been found as mummified offerings to solar deities, possibly in lieu of shrews when those animals were unavailable. **MYTHOLOGY/ICONOGRAPHY** While mice were thought of as vermin to be disposed of or trapped (a mouse trap from the town of Kahun/Lahun (UC 16773) survives in the Petrie Museum of Archaeology at University College London). In the New Kingdom a genie with a mouse head is sometimes shown introducing the deceased to the god Khentiamentiu, in his guise of Master of the West. Different sorts of mice feature in ostraca and Satirical Papyri, which depict topsy-turvy situations such as cats being subservient to mice, which are a form of social commentary. **HIGHLIGHTS** Ostraca and Satirical Papyri in many museums. Nile rats, gerbils and jirds are depicted on the Causeway of Unas at Saqqara, and in some Old Kingdom tombs such as that of Ptahhotep. A cat facing what is probably a jird can be seen in the tomb of Baqet III at Beni Hasan, and a jird features in the tomb of Ukh-hotep at Meir. A few statuettes of rodents are in museums such as the Egyptian Museum and the Metropolitan Museum of Art.

S. Ikram

Frequently mice were featured in images showing a topsy-turvy world, such as on this ostracon where a cat fans a mouse and offers it a duck, Brooklyn Museum (37.51E)

Common Ostrich ■ *Struthio camelus* Ht to 274cm WT to 154kg

DESCRIPTION Enormous flightless bird with long legs and long, slender neck. Both sexes huge with long, muscular legs, pinkish to pinkish-grey and with just two toes, the inner one being the larger, and a long, slender neck, pinkish to pinkish-grey and thinly covered with down. Head small with proportionately large eyes and short, flattened and rather pointed bill. Male has slate-grey to black body, loose, shaggy plumage with white primaries (functionless as flight feathers) and a bushy white tail. Females similarly loose plumaged, buff-brown throughout with young males similar. Chicks very different, being cream boldly marked with dark spots. **VOICE** Not vocal. Dull 'roar' recorded from male, mainly at night. **DISTRIBUTION** In Egypt a former breeder; status now uncertain and with just a few recent records right in the south of the country. Formerly over much of the Eastern and Western Deserts to Sinai. Extinct elsewhere in the region due to excessive hunting, but once occurred over Arabia to Jordan, Syria and Iraq. Now confined to sub-Saharan Africa, possibly north to Morocco. **HABITAT AND HABITS** Open plains and savannah grassland to semi-desert. Largely solitary but in breeding season may gather in larger groups. Flightless but has powerful legs and is a very fast runner, reaching more than 70km/ph. Male mates with several females after a spectacular courtship display, flapping the plumed wings. Eggs laid in a communal nest with a 'dominant' female. **SIMILAR SPECIES** None in Egypt. Records of ostriches from modern Egypt are of the Common Ostrich S. *camelus*, but some recent authors separate the race from Northern Kenya, Somalia and southern Ethiopia as the Somali Ostrich S. *molybdophanes*, in which the neck and legs of the male are more blue and brighter.

B. & T. O'Kan

Ostrich 𓅸 *New*

S. Ikram

Tutankhamun's Painted Box (Carter 021, JE 61467)

The largest of Egyptian birds, ostriches and their by-products played a key role in Egyptian culture and art. **PRACTICAL USES** Ostriches could be hunted and quite possibly bred. They provided meat to eat and feathers for adornment and fans. Their eggs were eaten and their shells provided the raw material for beads and vessels. Ostrich fat was valued as a food as well as in medicines. The skin could have been used as leather, although no examples of such a use survive. Their bones could be used to make tools or weapons. **MYTHOLOGY/ ICONOGRAPHY** Ostriches play no specific role in the mythology of ancient Egypt, although hunting them, especially by royalty, was part of the trope of the king's control over chaotic and wild nature. **HIGHLIGHTS** Ostriches are one of the few birds frequently featuring in rock art. Ostrich hunt scenes occur in tombs, starting in the Middle Kingdom in the tombs of Beni Hasan (Baqet III, No. 15) and Thebes, where in some instances they are brought back alive (Amenemhat TT 53, User TT 56, Rekhmire TT 100). An ostrich feather fan with some of the feathers still present survives from the tomb of Tutankham (Carter No. 272a, JE 61999). Ostrich egg vessels are displayed in the Egyptian Museum and the Grand Egyptian Museum.

Mute Swan ■ *Cygnus olor* L 140–160cm WS 200–240cm

DESCRIPTION Very large, all-white relative of the ducks and geese. Sexes similar. Adult white throughout. Bill deep orange-red with a small black tip and black basal margins running to eye. Bill knob black and prominent in male, less so in female. Legs relatively short and black. In flight note pure white throughout and neck extended. On water neck frequently held in characteristic 'S' shape and wings often held fluffed up, creating a domed body shape. Juvenile brownish-grey throughout with beige-grey bill. **VOICE** Not mute but with a honking call described as *heeorr*. Hisses when threatened. **DISTRIBUTION** In Egypt status uncertain and only recorded recently as a winter visitor. Widespread resident from Western Europe east through Eurasia, dispersing in winter. Widely introduced elsewhere. **HABITAT AND HABITS** Open fresh water on lakes and rivers generally with reedbeds; also coastal. Where feral, introduced on lakes and large ponds in parks and gardens. Dabbler, feeding by upending for water weed, when may reveal black legs. Monogamous, breeding on a large platform of reeds and grass, and aggressively defends nest. The Mute Swan has been introduced worldwide as an ornamental species and has established itself from these populations in many countries, such as Australia, New Zealand and South Africa. This has not happened in Egypt and it remains a very rare visitor. **SIMILAR SPECIES** The Mute Swan is one of the largest and heaviest flying birds. Other large, all-white birds, such as egrets and the Eurasian Spoonbill *Platalea leucorodia*, all have long legs and bills, and egrets fly with neck retracted. The Whooper Swan *C. cygnus* is an even rarer accidental visitor to Egypt, distinguished by black and yellow bill with no knob and neck held straight. Sits lower in the water without the fluffed up wings.

Tracy Robinson

Mute Swan *Akhy*

Only one uncontested image of a Mute Swan has been recorded from ancient Egypt, although other swans feature more commonly. **PRACTICAL USES** Food, feathers and bones could all be used. **MYTHOLOGY/ ICONOGRAPHY** The fact that statues of Mute (and possibly other) swans were placed in the Middle Kingdom burials of princesses Itiwert, Khnemt and Sit-Hathor-Meret in Dahshur, as well as in the Valley of the Kings tombs of Thutmose III (KV 34), Tutankhamun (KV62) and Horemheb (KV57) suggests that this bird had religious significance. **HIGHLIGHTS** A wooden statue of a Mute Swan in the tomb of the princess Itiwert from Dahshur in the Egyptian Museum. Images of swans appear in some of the mastabas at Saqqara, notably in the tomb of Ptahhotep II (D 64), where the accompanying text relates that he owned 1,225 swans.

S. Ikram

Swan as an offering; Tomb of Ptahhotep

H. Burton

Wooden swan covered with a resinous material from the tomb of Tutankhamun (KV62). Reproduced with permission of the Griffith Institute, University of Oxford

True Geese ■ *Anser* spp. L 56–88cm WS 115–174cm, and *Branta* spp. L 54–70cm WS 105–142cm

CC Dooncoombez

Greylag Goose

DESCRIPTION True geese are generally larger and longer necked than the shelducks and ducks. Most *Anser* geese are larger and longer billed than *Branta* geese and generally brown to grey-brown with bill colour important in distinguishing species. The Bean Goose A. *fabalis* has an orange patch, while the bill of the Greylag Goose A. *anser* is entirely orange-pink. In the White-fronted Goose A. *albifrons* and Lesser White-fronted Goose A. *erythropus* there is a white margin to the bill, in the latter extending on to the crown. The *Branta* geese are generally smaller, smaller billed and stockier, and more grey, black and white, often strongly patterned as in the Brent Goose B. *bernicla* and Barnacle Goose B. *leucopsis*. The Red-breasted Goose B. *ruficollis* has deep chestnut-red on the head, neck and breast. **DISTRIBUTION** All wild geese species are very rare in Egypt, with very few recent records, mostly in winter. Three goose species are portrayed in one of the most iconic paintings from ancient Egypt – the Meidum Geese (see opposite). Many species found globally but largely absent from sub-Saharan Africa. **HABITATS AND HABITS** Large, shallow inland lakes and lagoons. In winter also brackish lakes and estuaries but most historical records in Egypt have been from the Nile and the Delta Lakes. Geese generally graze. In flight flocks form distinctive skeins. **SIMILAR SPECIES** Any goose seen in Egypt should be assumed to be a Domestic Goose A. *a. domesticus*, which is derived from the wild Greylag Goose and can look very similar. Domestic Geese are generally heavier and are often entirely white or piebald or with white patches, for example on the wings. A further species, the Spur-winged Goose *Plectropterus gambensis*, a large black and white waterfowl with a bare, pinkish-red face has been recorded historically.

CC Johnny-on

Bean Goose

Geese *Semn*

Many types of geese existed in Egypt, some as residents, others as migrants. The latter could be captured and bred in captivity. It is possible to identify these birds to species with any confidence only when the paint survives to delineate their plumage. **PRACTICAL USES** As food, and the feathers were used to stuff pillows, and in decoration, in fans and in arrows. The fat was used for cooking and in medicines. **MYTHOLOGY/ICONOGRAPHY** While the White-fronted Goose seems to have been associated with the god Geb, it is not clear which goose was affiliated with Amun – possibly all geese, or maybe the Egyptian Goose. **HIGHLIGHTS** A row of geese is featured in the panel of the Meidum Geese in the Egyptian Museum. The many marsh and offering scenes found in the tombs of Saqqara and Giza show geese, while the tombs of Ti, Kagemni and Mereruka show different types of these birds being kept in captivity and force-fed. The tombs of Beni Hasan and Meir, and many tombs in Luxor, also feature the birds in marsh, fowling and offering scenes, and on the prows of boats or under chairs as pets. Victual or food mummies of geese are in the Animal Mummy Room of the Egyptian Museum, and the coffinets for mummified geese to be consumed as food from the tomb of Tutankhamun are kept in the Grand Egyptian Museum.

S. Ikram

Geese; Tomb of Ptahhotep, Saqqara

S. Ikram

The Meidum Geese

Egyptian Goose ■ *Alopochen aegyptiaca* L 70cm WS 144cm

B. Grewal

DESCRIPTION Small goose with very distinctive patterning. Largely buff-brown, darker on the back and with pale grey underparts, Buff on breast and with dark breast-spot. Pale grey head and neck with narrow dark collar, dark brown patch around eye and whitish crown. Tail black. Legs dark. Bill dull pinkish-red with dark basal margin and black tip. Legs relatively long and reddish-pink. Sexes similar. In flight wings dark with large white panels and a dark green speculum. **VOICE** Harsh braying *haah haah haah* and hissing. **DISTRIBUTION** In Egypt, common resident in the Nile Valley south from Aswan sometimes in large flocks, especially in winter, when it disperses further north along the Nile. Resident breeder over much of sub-Saharan Africa. Introduced and as an escape elsewhere, including in Britain, Germany, France, Spain and elsewhere in Europe, and with a feral population in the UAE. **HABITAT AND HABITS** Large inland waters, lakes and rivers, including the Nile at Aswan and the shores of Lake Nasser. Grazes on fields and grassland. Nest of grass twigs, and so on, in dense vegetation or tree hole. Territorial during breeding season and aggressive towards other species. **SIMILAR SPECIES** The Egyptian Goose is placed in the monotypic genus *Alopochen*. It seems most closely related to the shelducks, of which two species occur in Egypt. The Common Shelduck *Tadorna tadorna* is largely black and white with a chestnut breast-band and a bottle-green head. The bill is bright red with a prominent swollen base in the male. It is uncommon on passage and in winter on lakes and rivers. The Ruddy Shelduck *T. ferruginea* is warm orange-brown throughout, with a paler neck and head and, in the male, a narrow black collar. Bill and legs are black. In flight the wings are dark with a white wing-panel. In Egypt it is a rare winter visitor and migrant.

Egyptian Goose no specific name, but *Geb* or *Er*

Commonly pictured in marsh scenes and sometimes on the prows of boats and under chairs. When the paint does not survive it is difficult to identify these birds to species, as many geese and ducks made their home in Egypt, either permanently or transitorily. **PRACTICAL USES** As food, possibly as pets and as decoys when fowling. **MYTHOLOGY/ ICONOGRAPHY** Geese in general were associated with the god Amun, king of the gods, as well as Geb, the earth god, who also enjoyed the epithet of 'The Great Cackler'. In one creation myth Geb is said to have laid the egg out of which the sun hatched and life emerged. **HIGHLIGHTS** Nesting in a papyrus swamp with wings protectively outstretched in the Sixth Dynasty mastaba of Mehu at Saqqara, on the prow of a boat in the reliefs from the tomb of Nebamun in the British Museum, and possibly among food mummies.

S. Ikram

Tomb of Nebamun, British Museum (EA 37977)

Northern Pintail ■ *Anas acuta* 56cm (excluding tail feathers of male 10cm) WS 85cm

DESCRIPTION Slim, elegant, long-necked duck. Male greyish above with white breast and underparts, and black rear. Head and neck chocolate-brown with white from behind eye down side of neck. Long, slender, pointed tail diagnostic. Bill slender, dark grey. Female as the Mallard *Anas platyrhynchos*, but paler and slimmer with a plain head and pointed tail. In flight shows dark green speculum (duller brown in female) with white trailing edge. **VOICE** Quacking, as Mallard. Male also has a clear whistle. **DISTRIBUTION** Throughout Egypt and the region, a widespread winter visitor and also on passage. Breeds across northern Europe, Eurasia and North America. Winters further south to the Equator. **HABITATS AND HABITS** Large, shallow inland lakes and lagoons. In winter also brackish lakes and estuaries. In mixed species flocks note long neck, rather small head and pointed tail. **SIMILAR SPECIES** A number of other duck species occur in Egypt especially in winter and on passage and some of these were recorded with great accuracy in ancient Egypt. The Mallard is the wild ancestor of the domestic duck, to which other species can be compared. Male dark grey above, paler grey below with purple-brown breast and clear-cut black rear. Head bottle-green with thin white collar. Tail white. Bill yellow. Female mottled brown throughout with darker crown and eye-stripe. The Eurasian Teal *Anas crecca* is a small dabbling duck, the male largely grey with a white stripe along the side and a black-bordered, buff rear with the head dark chestnut with glossy dark green around and behind eye edged with yellow. Female like a very small Mallard. The Eurasian Wigeon *A. penelope* is a rather small, rounded duck. Male largely grey above and on flanks. Breast pinkish-brown and belly white. Head rounded, chestnut-brown with forehead to crown pale buff. Bill small, pale grey with black tip. Female darkish grey-brown, more reddish below with white belly.

S. Fonseca

Pintail Duck *Set*

Egypt was rich in waterfowl, particularly in ducks of all types, with the Pintail being very common. **PRACTICAL USES** The flesh was eaten, and the fat was used for cooking, cosmetics and medicine. The feathers could be used for decoration and the down for stuffing pillows. **MYTHOLOGY/ICONOGRAPHY** Ducks, as a whole, were regarded as wild creatures to be subdued in order to maintain order (*maat*), and scenes showing them being caught in clap-nets can be interpreted both literally as well as metaphorically, with the captured birds being equated with the enemies of Egypt. Some scholars have posited that ducks had erotic connotations. **HIGHLIGHTS** Both tombs and temples contain a plethora of images of different types of duck, many of which can be identified to species when the paint is present. The birds appear in marsh, hunt, offering and poultry yard scenes, as well as on offering tables. Cosmetic containers found in museums throughout Egypt and elsewhere often take the form of a duck. Mummified ducks, prepared for consumption, were found in the tombs of many people, notably that of Yuya and Thuju (Egyptian Museum) and Maherpri (Egyptian Museum). The coffinets containing Tutankhamun's duck mummies are in the Grand Egyptian Museum.

A. Oliveira

Ducks in a swamp; Tomb of Ay, Luxor

S. Ikram

Ducks and other waterfowl being hunted; Tomb of Menna, Luxor

Quail ■ *Coturnix coturnix* L 17cm WS 32–35cm

S. Fonseca

DESCRIPTION Very small, cryptically coloured game bird. Above buff-brown mottled with darker brown and streaked pale cream. Below has cream belly and undertail-coverts. Head complexly patterned with cream supercilium and dark through eye. Dark malar stripe. Throat variably dark in male, pale in female. Bill short and rather thin. In flight long winged and short tailed. When flushed flies fast and low, soon dropping back into cover. **VOICE** When flushed a soft *were*. Song a loud *pil pil-it* popularly rendered as 'wet-my-lips'. **DISTRIBUTION** In Egypt widespread passage migrant, especially along the North Coast and North Sinai. A few birds winter and an occasional breeder. Widespread throughout the region on passage and localized breeder. Elsewhere a widespread migrant breeder over much of west and south Eurasia, wintering in Africa (where also resident populations) and South Asia. **HABITAT AND HABITS** Open farmland, grassland and meadows. On passage any area of scrub and pasture, thickets and so on. Very shy and most often located when flushed (note small size and long wings) or by call. **SIMILAR SPECIES** The best chance of finding Quails in Egypt is along the North Coast in autumn, where the birds migrating from breeding grounds in Europe fly in off the Mediterranean into any area of coastal scrub and vegetation. Sadly, at this time of year vast stretches of nets are set up along the coast to net Quails. While the extent, spacing and positioning of the nets is all governed by legislation, in practice the nets extend uninterrupted along vast areas of coast. These nets are indiscriminate and catch huge numbers of other birds, especially passerines that become bycatch. Another victim is the Corncrake *Crex crex*, which looks superficially like the unrelated Quail but is slimmer, longer necked and with bright chestnut on the wing. It is threatened in Europe by changing agricultural practices and is a source of conservation concern.

Quail *Pa-at*

Quail chicks are common in Egyptian texts as they were the symbol for 'w', while adult birds often feature in Egyptian art. **PRACTICAL USES** Quails were eaten, as were their eggs. **MYTHOLOGY/ICONOGRAPHY** Quail chicks represent 'w', and the adult birds are part of the usual corpus of wild birds that need to be controlled so that order (*maat*) triumphs over chaos (*isfet*). **HIGHLIGHTS** Numerous texts include quail chicks. The adult birds are shown in many Memphite tombs; for example, they are netted in Mereruka's mastaba at Saqqara, as well as in Theban tombs, for example that of Khaemhet (TT 57). They also appear, often tied to sheaves of wheat, for example in the Theban Tombs of Ramose (TT 55), Menna (TT 69), and Nakht (TT 52), in offering processions on both tomb and temple walls.

S. Ikram

Right: Quails and wheat; Tomb of Ramose (TT55), Luxor

S. Ikram

Quail chick; Tomb of Kagemni, Saqqara

J. M. Galán

Painted quail chick; Coffin of Iqer

Great Cormorant ■ *Phalacrocorax carbo* L 85cm WS 140cm

Tracy Robinson

DESCRIPTION Large, long-tailed, all-dark waterbird. Adult dark, oily black glossed blue-green and bronze. Cheeks and chin white with yellow chin-patch, short, angular crest and white flank-patch. Bill long and relatively powerful, with hooked tip. Legs and feet dark. In flight, long, rather thick neck extended and kinked. Note long tail. Duller in winter and immature pale below. **VOICE** Guttural *kaarks* but generally silent outside breeding season. **DISTRIBUTION** In Egypt, widespread passage migrant and winter visitor. Occurs on all coasts and inland waters throughout Delta and Valley, including Lake Nasser. In region occurs throughout in winter and on passage. Breeds over much of Europe, east through Asia and to Australia; also eastern USA. **HABITAT AND HABITS** Coastlines including lagoons and offshore islands; also large inland lakes and rivers, and in Egypt along the Nile including central Cairo. Large roost in Aswan at the Botanical Gardens. Also roosts on sandbanks and rocks and cliffs. Large winter flocks on Lake Nasser. Roosts in large trees. On water sits very low with bill held tilted up. Dives for fish with short half-leap. Flies in lines and 'V's. **SIMILAR SPECIES** The Shag *P. aristotelis* is a smaller, slimmer version of the Great Cormorant, distinguished by the slimmer, all-dark bill, lack of white on the cheek or flanks and with a prominent crest. It is very rare in Egypt. The African Darter *Anhinga rufa* is related to the cormorants but is much slimmer, with a longer, more slender neck with a pronounced kink. It is largely blackish with white streaks and with a dark rufous neck and white neck-stripe. The bill is long and dagger-like, lacking the hook tip. There are no records from modern Egypt but it has been identified from pharaonic friezes (see opposite) and could conceivably creep into Egypt on Lake Nasser.

Cormorant 𓅧𓈎 *Aq*

Cormorants or birds that look like them are frequently featured in Egyptian art from the Old Kingdom onwards. Sometimes they are shown perched on papyrus umbels, which is possible, but it is probably not their preferred position. The word for cormorant, as well as the bilateral sign, is *aq*, evocative of their cry during mating. **PRACTICAL USES** Probably as food. There is no incontrovertible evidence for these birds having been used to help in fishing as they have been used traditionally in China and Japan, as well as in Greece and Northern Macedonia, notably on Doiran Lake. **MYTHOLOGY/ICONOGRAPHY** Not related to any deity, but features in swamp scenes and as part of the hieroglyphic writing system. **HIGHLIGHTS** In the swamp scenes in Old Kingdom tombs, notably that of Mereruka at Saqqara, as well as swamp scenes of the Middle Kingdom, such as the tomb of Khnumhotep III (Tomb 3) at Beni Hasan. The darter is pictured in the Botanical Garden of Thutmose III at Karnak Temple. This 'garden', carved in stone, recorded exotic creatures that were brought into Egypt from the Levant by King Thutmose III and were presumably kept in the temple's gardens and menageries.

S. Ikram

A cormorant eating a fish; Temple of Hatshepsut, Deir el-Bahari

Black-crowned Night Heron ■ *Nycticorax nycticorax* L 62cm WS 94cm

DESCRIPTION Medium-sized, stocky heron with heavy bill. Adult largely pale grey to whitish below with grey wings. Mantle, back and crown black. White at base of bill. Two elongated, slender plumes from crown not always distinct. Bill black and moderately heavy. Iris red – good identification feature at close range. Legs and feet yellowish, turning redder when breeding. In flight, note compact build and uniform pale grey wings. Juvenile very distinct, being pale brown throughout with clear pale spotting on upperparts, more streaked below. Bill yellow-brown with darker tip. **VOICE** Call a croaking *quark*. Vocal in breeding colonies. **DISTRIBUTION** In Egypt, widespread in winter and on passage throughout the Delta and Valley. Probable resident breeder in southern Valley and perhaps Delta. Widespread on passage and in winter throughout region, with localized breeding in the Levant and eastern Arabia. Elsewhere across South and South-east Asia, sub-Saharan Africa, southern Europe and much of the Americas. **HABITAT AND HABITS** Wetlands including lakes, rivers and marshes. Mainly nocturnal, roosting by day in trees and dense cover where unobtrusive. Best noted at dusk, when flocks fly out from roosts to feed, where stocky build, stout bill and stiff wingbeats mark them out from other heron and egret species. **SIMILAR SPECIES** The much smaller Little Bittern *Ixobrychus minutus* is just 35cm long. The male shares the black crown and mantle, but is otherwise very pale buff and in flight with much black on the wings. The female is more streaked, and brown, streaked black on the upperparts. The Striated Heron *Butorides striata* is a smaller, darker species than the Black-crowned Night Heron. It is largely blue-grey to brown-grey above with beige feather fringes above, more uniform below and with a grey neck. The crown is black and with a black cheek-stripe. The iris is yellow. Formerly limited to the Red Sea coasts and South Sinai, this species has now expanded its range to the Nile Valley north to Cairo.

T. Robinson

Black-crowned Night Heron *Benu*

While this bird is well known in Egypt, it is rarely depicted in Egyptian art, or at least, it is not easily recognizable as such without paint. **PRACTICAL USES** Possibly used as a decoy to trap birds (tomb of Kagemni, below), which suggests taming. **MYTHOLOGY/ ICONOGRAPHY** As birds are part of nature that needs to be controlled so that order may rule, this bird is also possibly shown in such a context. Also, as it is a nocturnal creature it might be associated with the nocturnal aspects of the sun god. **HIGHLIGHTS** The only clearly identifiable examples of this bird are found in the Eleventh Dynasty tomb of Baqet III (No. 15) at Beni Hasan, and the mastaba of Kagemni at Saqqara, where it is shown near a fowlers' clap-net.

S. Ikram

Herons in the marshes; Tomb of Mereruka, Saqqara

Little Egret ■ *Egretta garzetta* L 60cm WS 90cm

DESCRIPTION Medium-sized, elegant, all-white egret. Both sexes white throughout. Lores blue-grey for most of the year, turning reddish-pink during courtship. In breeding plumage there are gauzy plumes over the back and two long white, narrow plumes from back of head. Bill long, straight and slender, generally blackish. Legs black with sharply contrasting, bright yellow feet. In flight all white, neck retracted and feet extending beyond tail. **VOICE** Generally silent outside breeding but call note a rather harsh *aaagh*. **DISTRIBUTION** In Egypt widespread passage migrant and winter visitor. Also, local resident breeder in the Delta and Valley, for example at Bilbais and Aswan. In region occurs throughout in winter and on passage, and is a local breeder in the northern Levant. Breeds over much of southern Europe, east through southern Asia to Australia, and also sub-Saharan Africa. **HABITAT AND HABITS** Wetland areas including lakes, rivers, marshes and irrigation canals, and also on coast, where it needs to be distinguished from the Western Reef Heron *E. gularis* by bill colour and black legs and yellow feet. Breeds in colonies, often with other heron and egret species. Stalks in shallow water, lunging and stabbing at prey. **SIMILAR SPECIES** Other all-white egret species are found in Egypt. The Great Egret *E. alba* is a common winter visitor to the Delta and Valley. It is much larger and in non-breeding birds bill is yellow. Legs proportionally longer and lack contrasting yellow feet. The Cattle Egret *Bubulcus ibis* is a small, stocky egret with a relatively short, yellow-grey bill and yellowish legs. In summer the all-white plumage is relieved by buff-orange on the head and breast. This is the common egret of the farmland, attending sheep and cattle in the fields, and in towns and villages scavenging on garbage piles.

H. Çaglar Güngor

Little Egret *Seda*

A variety of egret features in Egyptian art. **PRACTICAL USES** Possibly used as decoys by fowlers, as food, feathers and bones used to make tools, weapons or personal adornment. **MYTHOLOGY/ICONOGRAPHY** Images of egrets appear in marsh scenes and they are also shown being used as decoys – indeed, until relatively recently Egyptians (and others) have used egrets as decoys and also tamed them. Cattle Egrets are known in modern Egypt as the friend of the farmer as they accompany farmers while they till the soil, eating insects that might threaten the crops. **HIGHLIGHTS** Many Memphite mastabas contain images of egrets in marsh scenes. Egrets used as decoys are common in Theban tombs, such as that of Userhat (TT 56) and Menna (TT 69).

S. Ikram

Egrets being used as decoys; Tomb of Menna (TT69)

S. Ikram

Fowling in the marshes with egrets; Khnumhotep II, Beni Hasan Tomb 3

Sacred Ibis ■ *Threskiornis aethiopicus* 80cm WS 112cm

Bernard & Tara O'Kane

DESCRIPTION Large, largely white bird with black head and neck. Body white with naked black head and long neck, and black scapulars spilling over rump and tail. Bill long, black and downcurved. Legs black. In flight largely white with black margins to primaries and secondaries. Flies with neck extended (note that herons fly with neck retracted). **VOICE** Silent except when breeding, when high 'yelps' and grunts recorded. **DISTRIBUTION** Extinct in Egypt, with the last accepted record being from 1891. Widespread throughout sub-Saharan Africa north to Sudan, and in south-east Iraq. Introduced elsewhere, including in France, and feral populations in the Arabian Gulf. Two similar species across South Asia and Australasia sometimes considered conspecific. **HABITAT AND HABITS** Marshes, swamps and similar wetlands, as well as parks, large gardens and farmland. Nests in colonies. **SIMILAR SPECIES** The Sacred Ibis has been extinct in modern Egypt for more than a century. With the creation of Lake Nasser and potentially new wetland habitat the species may recolonize Egypt in the future. There is also discussion over the reintroduction of the species into Egypt from the Sudanese population. As yet this has not happened. Two other ibis species have been recorded from modern Egypt. The Northern Bald Ibis *Geronticus eremita* has been recorded but with the latest confirmed record from 1962. It is now Critically Endangered. There have been two records from the 2020s, both in southern Sinai, but they have yet to be confirmed by the Egyptian Ornithological Rarities Committee (EORC). The Glossy Ibis *Plegadis falcinellus* is a migrant and winter visitor, and possible breeder. It has all-dark, glossed green-purple plumage in good light, with a long neck and legs, and a long, downcurved bill. Look for it throughout Egypt along riverbanks, marshes and wetlands. It often occurs in flocks, and in flight note the outstretched neck and tendency to fly in lines.

Sacred Ibis *Heby*

A bird that is found in both sacred and secular contexts, sometimes with the Glossy Ibis. **PRACTICAL USES** As food, as mummified offerings, in medical preparations and as a component of magical rituals. **MYTHOLOGY/ ICONOGRAPHY** The Sacred Ibis was identified with Thoth, the god of writing and wisdom (his other totemic animal was the baboon). The ibis was connected with Thoth as its beak resembles a pen, as well as curving to evoke the crescent moon, reflecting Thoth's lunar character, further emphasized by the bird's white plumage. The fact that the ibis constantly ferrets about with its beak also suggested its quest for knowledge. In addition, it is possible that the fact that ibis eat snails, the host for the parasites that cause schistosomiasis, made the Egyptians equate the bird with Thoth, who also had curative powers. However, there is limited evidence that ibis did eat snails of the genera *Biomphalaria* (*B. alexandrina*) and *Bulinus*, both of which are transmitters of the disease. **HIGHLIGHTS** There are a myriad images of the Sacred Ibis in texts, on tomb and temple walls, and in the form of animal mummies that were offered in the hundreds of thousands to Thoth. Marsh scenes in the Memphite and Theban necropoleis boast images of the Sacred Ibis, as well as the Glossy Ibis *Plegadis falcinellus*. Amulets of Thoth as an ibis, rendered in all sorts of materials, abound in Egyptian museums, as well as museums throughout the world, as do statues of this bird. In some cases, the three-dimensional images were actually coffins for ibis mummies (examples in the Egyptian Museum, Mallawi Museum and Grand Egyptian Museum). Modern Ashmunein was dedicated to Thoth, and the nearby necropolis of Tuna el-Gebel contains many mummified ibises, as do Saqqara and Abydos. Images of Northern Bald Ibises are rare, but exist as hieroglyphs and sometimes adorn jewellery, notably diadems of the Old Kingdom.

A.-M. Kellen

Well-wrapped ibis mummy; Egyptian Museum (CG 29864)

A. Gabr

Sacred Ibis in the marsh; funerary complex of king Userkaf, Egyptian Museum

Greater Flamingo ■ *Phoenicopterus roseus* L 120–145cm WS 140–170cm

T. Robinson

DESCRIPTION Large, tall, slender and unmistakable water bird with very long neck and legs. Adult white to pinkish-white throughout, with deeper pink on wings. Bill deep pink with black tip and sharply angled down. Legs very long, slender and bright pink. In flight, note deep pink-red on wings with black flight feathers. Flies with very long, slender neck extended. Juvenile brownish-grey throughout with beige-grey bill and brownish legs. **VOICE** Honking and braying, especially in flight and in flocks and breeding colonies. **DISTRIBUTION** In Egypt former breeder with evidence that breeding may have been resumed on Lake Qarun, Fayoum. Otherwise regular winter visitor in singles and flocks, sometimes numbering hundreds, on shallow lakes in northern Delta, Fayoum and Siwa. Occasional on coast. Widespread in winter and on passage in region and very localized breeder; also sub-Saharan Africa and western Asia to India. **HABITAT AND HABITS** Shallow lakes, salt flats and coastal lagoons. Generally in flocks, but also singles. Feeds by dipping the angled bill into the water and filtering out small crustaceans through the tongue. **SIMILAR SPECIES** Unmistakable given good views, but from a distance can appear all white until it takes flight. The Eurasian Spoonbill *Platalea leucorodia* (L 90cm) is also long legged and long necked, occurs in similar habitat and can be confused at a distance. It is all white with dark legs and a long, straight, spatulate bill, black with a yellow tip. Bill pale in juvenile. In flight there is no pinkish-red or black on uniform white wings. Local winter visitor to the Delta, Northern Lakes and Valley, and very local breeder on islands along Red Sea coast.

Greater Flamingo *Desher*

The word for flamingo, *desher*, was used not only for the bird itself, but also for the colour red and its derivatives. *Deshret*, the word for desert, was also derived from this root, as some of the sands could be a fiery red. **PRACTICAL USES** Possibly eaten, but not commonly. Feathers may have been used in headdresses or ornamentation, but none has been positively identified in the archaeological record. **MYTHOLOGY/ICONOGRAPHY** None, though possibly, due to its colour, associated with the sun, blood and power. **HIGHLIGHTS** Flamingoes feature on pottery from the Naqada II period, and are frequently shown in marsh scenes, particularly in Old Kingdom tombs in Saqqara, as well as being used in hieroglyphic texts.

Courtesy of the Metropolitan Museum of Art

Box with flamingoes, Metropolitan Museum of Art (10.130.1175)

S. Ikram

Tomb of Mereruka, Saqqara

Egyptian Vulture ■ *Neophron perenopterus* L 55–65cm WS 155–170cm

DESCRIPTION Medium-large raptor though small for a vulture, thin billed, and black and white. Adult largely white tinged grey or brownish, with black flight feathers. Head rather small with bare, bright yellow facial skin and shaggy, whitish crest. Bill slender and yellow, with dark hooked tip. Immature uniformly brown with grey-blue facial skin. Matures over five years, with brown gradually replaced by white. In flight, note long, wedge-shaped tail and wings with relatively deep 'fingers'. In adult, from below secondaries and primaries black. From above flight feathers with white centres. **VOICE** Generally silent. **DISTRIBUTION** In Egypt mainly on passage but a scarce breeder, with nesting recorded in the Valley, Eastern Desert and Sinai. In region, summer breeder in the Levant and Naqab and resident in Arabia. Has declined virtually everywhere. In Egypt best seen in spring at Ain Suhkna using the thermals over the cliffs to gain height. **HABITAT AND HABITS** Ranges over mountains, wadis and gorges in semi-desert country and also attracted to rubbish tips around towns and villages. Feeds largely on carrion and offal. Breeds on cliff ledges. Generally solitary but gathers at carcasses. With other vulture species at carcasses, generally last to feed, using thin bill to take smaller morsels. **SIMILAR SPECIES** The Egyptian Vulture is also known as the Pharaoh's chicken and is sufficiently different from other vultures to be placed as the sole member of the genus *Neophron*. It is one of the very few birds to use tools. Egyptian Vultures use stones to open Ostrich eggs, swinging the stone repeatedly down on the egg until it is smashed. This has never been historically noted locally, and the Ostrich is now possibly extinct in Egypt. While the Egyptian Vulture is unlikely to be mistaken for any other raptor in Egypt, the White Stork *Ciconia ciconia* is also large, all white and with black flight feathers. It, however, flies with the long, slender neck held outstretched, and has long red legs extending beyond the tail.

CC Koshy Koshy

Egyptian Vulture A or *Aas* or *Nerit*

The Egyptian Vulture is the first letter of the Egyptian alphabet, with the sound of 'a'. While common in texts, it is rarely shown in imagery. **PRACTICAL USES** Egyptian Vultures, now uncommon, used to be plentiful in earlier times. Indeed, in the nineteenth century and earlier, travellers dubbed them 'Pharaoh's chickens', as they were unafraid of humans and almost tame. These scavengers focused on rubbish dumps, feeding on offal, carrion and excrement, thus playing a key role in waste disposal. **MYTHOLOGY/ ICONOGRAPHY** None specific to it, but could be associated with the goddess Nekhbet, although the Griffon and Lappet-faced Vultures (p. 113) were more commonly affiliated with her. **HIGHLIGHTS** In the Egyptian Museum a painted image from the mastaba of Atet. A unique image is found in Theban Tomb (TT127) belonging to Senemiah. Here the bird is shown on the papyrus umbel in a papyrus thicket, but this is curious as the vulture is not normally found in this environment. Otherwise, it is commonly featured in texts.

S. Ikram

Tomb of Rekhmire (TT100), Luxor

Lappet-faced Vulture ■ *Torgos tracheliotus* L 100–112cm WS 259–290cm

Shutterstock/EcoPrint

DESCRIPTION Huge vulture. Adult enormous with uniform dark brown upperparts, paler below, pale streaked dark with pale flanks. Head and neck naked and largely pale pink. Neck ruff shaggy. Bill huge and powerful. Immature has uniformly dark underparts and more blue-grey head and neck than adult, and adult plumage is acquired gradually over 6–7 years. In flight, note huge wingspan and very broad wings with a slight bulge in the secondaries and deep 'fingers'. **VOICE** Generally silent. **DISTRIBUTION** In Egypt now restricted to the southern Eastern Desert, where it breeds. Occurs sparsely over the Arabian Peninsula in the centre and south. Isolated population in Naqab possibly extinct. Elsewhere sparsely distributed over much of Africa except some western and central parts. **HABITAT AND HABITS** Ranges over mountainous desert and semi-desert country, ravines, gorges and wadis. Breeds on mountain crags, with a huge nest of sticks and branches. Generally solitary but gathers at carcasses. Food predominantly carrion where species is dominant over other vulture species. **SIMILAR SPECIES** Two other similar vulture species occur in Egypt. The Griffon Vulture *Gyps fulvus* is an uncommon passage migrant in Egypt mainly in the east. It is slightly smaller than the Lappet-faced Vulture and paler above with a downy white head and neck and pale ruff. The Black Vulture *Aegypius monachus* is of similar size to the Lappet-faced Vulture but is all dark below with a pale brown ruff and pale, not pink, head and black throat. In flight there is no bulge on the trailing edge of the secondaries. It is very rare and irregular on passage and in Egypt. A third species, the Lammergeier or Bearded Vulture *Gypaetus barbatus*, is of similar size but differs radically in silhouette, with long, narrow wings and a long, wedge-shaped tail. It is a very rare breeder in the southeasternmost corner of the country.

Lappet-faced Vulture *Nerit?*

S. Ikram

Goddess Nekhbet; Temple of Seti I, Abydos

Sometimes in Egyptian art the Lappet-faced Vulture and Griffon Vulture are interchangeable in terms of iconography, although often attributes derived from both birds are represented, thereby combining the two. When a single vulture is depicted, the most commonly depicted one is the Griffon Vulture. **PRACTICAL USES** The feathers may have been used in religious garb and for fans. **MYTHOLOGY/ICONOGRAPHY** The vulture was associated with the goddess Nekhbet, the deity representing Upper Egypt, whose main cult centre was the site of el-Kab, south of Luxor. She was paired with Wadjet (see Cobra, p. 137), the goddess of Lower Egypt, and they figure together on the brows of kings, particularly from the reign of Tutankhamun and thereafter. The Griffon Vulture was also affiliated with the goddess Mut, whose name also means 'mother' in Egyptian. Mut was the consort of the god Amun, chief deity from the Twelfth Dynasty onwards; thus she features prominently in Theban temples, particularly in Karnak. Generally, Egyptian queens were identified with her and are depicted wearing a headdress consisting of a schematic vulture with the wings hanging down on either side of the head, the head over the queen's brow, and the tail hanging along the back of her head. **HIGHLIGHTS** Images of Nekhbet feature in royal jewellery, with an exquisitely crafted pectoral showing a vulture with its wings spread and clasping the *shen* sign in its talons found in the tomb of Tutankhamun. Vultures with outspread wings are depicted on the ceilings of royal tombs in the Valley of the Kings, as well as on the central aisles of temples. In many temples, vultures are shown hovering over the king, protecting him.

S. Ikram

Vultures on the ceiling of Dendara Temple

Common Kestrel ■ *Falco tinnunculus* L 31–37cm WS 68–78cm

DESCRIPTION Smallish falcon with long wings and tail. Sexes differ. Male has chestnut upperparts with small black spots. Underparts pale, spotted dark. Rump and tail grey with broad black terminal band. Head grey with darker moustachial stripe. Female larger, with brown head and brown tail with dark bars and terminal band. Both sexes have dark claws. In flight, note long tail and pointed wings, although they can appear broader tipped when spread. **VOICE** Urgent *kee kee kee kee …* especially in spring; also quavering trills. **DISTRIBUTION** In Egypt a widespread resident including in the North Coast, Delta, Valley, Western Oases and Sinai. In region widespread resident. Also found over much of Europe and South and East Asia. **HABITAT AND HABITS** Open country, although not true desert, farmland, villages, towns and cities including Cairo. Can be associated with human activity, often nesting on buildings, including monuments. Feeds on rodents, small birds, reptiles and large insects such as locusts and beetles. Often hunts by 'hovering', flying into the wind and hanging motionless in the air before diving down on to prey. **SIMILAR SPECIES** Eleven species of falcon have been recorded in Egypt, of which the Common Kestrel is the most widespread. The Lesser Kestrel *F. naumanni*, an uncommon migrant, is very similar, but the male has an unspotted mantle and plain grey cheeks. Female best distinguished by white claws. The resident Lanner Falcon *F. biarmicus* is up to 50cm long with white cheeks, buff crown and slender, dark moustachial stripe. The Peregrine Falcon *F. peregrinus* is similar but more powerfully built, darker and with a thicker moustachial stripe and no buff on the crown. It is a scarce winter visitor even to central Alexandria and Cairo. In the desert and on the shores of the Red Sea look out for the Sooty Falcon *F. concolor*. The adult is uniform sooty-grey throughout with yellow legs, cere and eye-ring. It times its breeding, for example at Wadi Gimal, to coincide with the autumn migration of passerines coming from breeding grounds in Europe.

T. Robinson

Kestrel Djeret

One of the many birds of prey occasionally featured in Egyptian art. **PRACTICAL USES** As votive animal mummies. **MYTHOLOGY/ICONOGRAPHY** In the Nineteenth Dynasty, kestrels are sometimes shown as the goddesses Isis and Nephthys, although generally they are shown as kites, who shriek loudly as if mourning the dead. It is difficult to differentiate between the various types of falcon that might have been depicted when the paint is missing, as size and pelage are the main differentiators. It should be noted that raptors (mainly falcons, hawks and eagles) are associated with the sun god, as they fly high in the sky, and have yellowish-gold plumage and bright eyes that resemble the sun. Also, sometimes different types of falcon are conflated to create a 'super-raptor', such as the god Horus. He is a solar god and the personification of the king on Earth, who is depicted using elements from Eleonora's Falcon *F. eleonorae*, the Hobby *F. subbuteo*, Lanner Falcon and Peregrine Falcon. Many kestrel mummies have been found at sites dedicated to Horus – possibly these are the most plentiful raptors offered, as kestrels are relatively easy to breed. **HIGHLIGHTS** The tomb of Sennedjem (TT1) in Deir el-Medina, on Luxor's west bank, contains images of the goddesses Isis and Nephthys as kestrels. The tomb of Ipi (TT217), also in Deir el-Medina, shows what might be a kestrel or other raptor waiting to seize meat and offal near a butcher's shop. The Horus Falcon is commonly shown as a bird (see the statue of King Khafre in the Egyptian Museum and images of the bird above the royal name), or as a man with a falcon head (in all Egyptian temples and many royal tombs in the Valley of the Kings).

K. Amaury

Sometimes kestrels were substituted for kites, though kites were the more common bird associated with both Isis and Nephthys as they mourned the god Osiris; Tomb of Nefertari, Luxor

Purple Swamphen ■ *Porphyrio porphyrio* L 48cm WS 90–100cm

(Purple Gallinule)

DESCRIPTION Spectacular but elusive large rail, superficially fowl-like. Sexes similar. Chunky, dark glossy purple-blue throughout, with race *madagascariensis* occurring in Egypt with greenish back. Undertail-coverts white. Head has red face shield and very stout, heavy red bill. Legs and very long toes red. In flight all dark with red bare parts and dangling legs and feet. **VOICE** Vocal. Varied repertoire of clucks, honks and brays; also loud *chuk* and shorter, sharper *tschak tschak*. **DISTRIBUTION** In Egypt resident breeder throughout the Delta and Valley, including Fayoum and the Northern Lakes. Also found in southern Europe, parts of the Middle East, north-west and sub-Saharan Africa and much of South and South-east Asia to Australasia, with races sometimes treated as distinct species. **HABITAT AND HABITS** Wetlands along lakes, rivers, irrigation canals and reedbeds with dense cover. Elusive but where undisturbed can be seen out in the open stalking in shallows. When disturbed makes for cover, when bright white undivided undertail-coverts striking. Easily seen on vegetated shores of Lake Qarun, Fayoum and Saluga and Ghazal Protected Area, Aswan. Often seen clambering in thick vegetation, where bright red bill and legs are distinctive. **SIMILAR SPECIES** The Moorhen *Gallinula chloropus*, at 31cm, is a smaller version of the Purple Swamphen, but dark blackish-slate throughout tinged brown on the back and with white undertail coverts and a streak along the flank. Legs greenish-yellow and face-plate and bill red, the latter with yellow tip. Common wetland resident and winter visitor throughout Egypt. The Eurasian Coot *Fulica atra* is superficially similar but slate black throughout with a white face-plate and bill. With good views, the greenish legs and feet show lobed toes. It is a localized resident and common winter visitor to wetland areas.

Purple Swamphen 𓊃𓎛𓎛𓏏𓅮 *Sehihet* and 𓅮𓅮𓅮 *Seshew* for swamp fowl

Bird that is commonly depicted in Nilotic scenes. **PRACTICAL USES** Eaten. Its feathers may have been used for decoration. **MYTHOLOGY/ICONOGRAPHY** These birds, like many others, embodied the marshy areas of Egypt and the wild creatures that needed to be controlled. **HIGHLIGHTS** Purple Swamphens are frequently depicted in marsh scenes found in the Memphite tombs – for example, the Giza tomb of Nebemakhet (LG86), and the Saqqara tomb of Niankhkhnum and Khnumhotep at Saqqara – as well as in royal contexts, such as the relief from King Userkaf's pyramid complex, now in the Egyptian Museum. Baqet III's tomb (No. 15) at Beni Hasan also contains an image of this bird.

A. Gabr

Purple Swamphen in a marsh; funerary complex of Userkaf

Common Crane ■ *Grus grus* 90–119cm WS 180–230cm

DESCRIPTION Large, stately waterbird. Superficially heron-like but larger and shorter billed, tall and elegant. Sexes similar. Largely grey throughout with elongated tertials forming a distinct, shaggy bustle at rest. Neck black at front to throat and face, white at back to behind eye, with bare patch of red skin on crown often indistinct. Bill horn coloured, longish and straight. Legs dark. In flight note wings uniform grey below, above with black primaries. Note deep 'fingers'. Flies with neck extended (herons fly with neck retracted). In migrating flocks often in lines or 'V' formation. **VOICE** Vocal. In flight contact call deep, sonorous and far carrying *krro krro*.... **DISTRIBUTION** In Egypt fairly common throughout on passage, and large flocks can be seen at migration hotspots such as Ain Suhkna and Suez. In region widespread on passage with very few wintering. Breeds over northern Eurasia, wintering in parts of southern Europe, Africa and South Asia. **HABITAT AND HABITS** Marshes, swamps and similar wetlands and farmland. On passage, migrating flocks often heard before seen. At Ain Suhkna in spring flocks use the thermals, rising against the cliffs to gain height ahead of continuing north. On the ground, elegant. **SIMILAR SPECIES** The Demoiselle Crane *Anthropoides virgo* is a slightly smaller (L c. 90–100cm) and much rarer species that occurs as a rare passage migrant in Egypt. It is pale grey throughout with an all-black head and neck, and white plumes curling down from behind the eye. The tertials are long and slender. There are recent records from Fayoum. Otherwise the cranes can be distinguished from the superficially similar herons and egrets by the shorter, less dagger-like bill, the elongated or bushy tertials and in flight by flying with the neck outstretched.

B. Grewal

Common Crane *Aw* or *Djat*

Cranes feature in scenes showing the riverine environment as well as poultry yards. **PRACTICAL USES** Cranes were delicacies to be eaten, both by humans and by the gods. Scenes show them being force-fed in order to make them tastier and plumper. The bones could be used to make tools or musical instruments, and the feathers might have been used for adornment. **MYTHOLOGY/ICONOGRAPHY** Cranes probably fell under the general rubric of wild creatures that had to be kept in check so that order rather than chaos would reign. **HIGHLIGHTS** Images of cranes, both the Demoiselle Crane *Grus Virgo* and Common Crane, are abundant in Old Kingdom mastabas, most notably those of Ti, Mereruka and Kagemni at Saqqara. They are shown being brought as offerings to the deceased, as well as being force-fed. Cranes appear in the New Kingdom temples in Thebes (Luxor), where they are often draped over offering tables. Some fine painted examples survive in Hatshepsut's temple at Deir el-Bahari.

S. Ikram

Force-feeding cranes; Tomb of Mereruka, Saqqara

S. Ikram

Group of cranes; Tomb of Ti, Saqqara

Northern Lapwing ■ *Vanellus vanellus* L 30cm WS 70cm (Peewit)

T. Robinson

DESCRIPTION Large, wispy-crested plover. Adult dark above with green-purple gloss; feathers fringed with pale buff in winter, giving a more scaled appearance. White below with broad, blackish chest-band. Vent orange-brown. Head black and white with steep forehead and in winter with white throat. Note the long, slender crest often held erect. Bill dark, short and straight. Legs orange-brown. In flight, note white rump with orange-brown on upper rump, dark, broad, rounded wings with white tips and white underwing-coverts. Tail white with broad black subterminal band. **VOICE** Vocal. Flight call an urgent *pee wich* often repeated. Known to call at night. **DISTRIBUTION** In Egypt widespread in winter and on passage throughout the Delta, including the Northern Lakes and Fayoum and Valley. Widespread on passage and in winter throughout region. Elsewhere a breeding resident in Western Europe and summer breeder over Eurasia, wintering further south to India. **HABITAT AND HABITS** Open farmland, wetlands including lakes, rivers and marshes; also coastal in winter. Frequently in flocks but also singly. Despite rather striking appearance, unobtrusive in dark vegetation. **SIMILAR SPECIES** The Spur-winged Lapwing *V. spinosus* is a related species, slightly smaller, with pale grey-brown upperparts, black hood and throat extending down breast to black belly, and bright white cheeks and sides of neck. Tail largely black with white rump. Legs rather long and black. In flight very black and white. Common and very noisy resident throughout the Delta, Valley and Fayoum on open farmland, marshes, irrigation canals, golf courses, hotel grounds (including Mena House, Cairo, Crocodile Island, Luxor and Movenpick, Aswan) and parks. Call a loud *pik pik*, especially urgent around nest site. Nests in scrape on open ground, where eggs highly camouflaged.

Lapwing Rekhyt

Commonly shown in a variety of contexts from the Predynastic through the Ptolemaic Period. **PRACTICAL USES** Pets, and possibly as food, in extremis. **MYTHOLOGY/ ICONOGRAPHY** The Lapwing is the name for the common people, the population of Egypt, probably because these birds are plentiful in Egypt, particularly in the Delta, and also because they utter a plaintive shriek, like a supplicant. Most commonly, the bird is somewhat anthropomorphized by being given arms that are raised in praise to the king. The image of a Lapwing with its arms raised, seated on a basket, with a star in front of the bird, is found in many temples beneath scenes showing the king, and can be read as: all (the basket) the common people (the Lapwing) praise (the star) the king (pictured in the register above). **HIGHLIGHTS** Like Hoopoes *Upupa epops* Lapwings are sometimes shown as children's pets, mainly in the hands of girls (for example in the tomb of Nefer at Saqqara, as well as other tombs there). They feature in marsh scenes in the Memphite area, as well as in Thebes (such as at Montuemhet's Theban Tomb, TT34, and Amenemhat's Theban Tomb, TT82). Lapwings are found on thrones, in temples, tombs and many royal contexts, as they are the symbol of the people of Egypt.

S. Ikram

The Lapwing representing the common people of Egypt, praising the name of king Ramesses II, Karnak Temple

European Turtle Dove ■ *Streptopelia turtur* L 27cm WS 47cm

CC C. J. Sharp

DESCRIPTION Rather small, slight, dark dove. Upperparts have tortoiseshell pattern with dark brown feathers with orange margins. Head pale grey with grey-pink face and breast becoming whitish below. Distinct black and white patch on neck-sides. Eye has pale orange iris, red orbital ring and bare skin. Legs red. In flight, note grey rump and grey tail with striking black and white borders, black to tip in centre. Wings dark above with pale grey panel and uniform grey below. **VOICE** Soft purring rendered as *turrrr turrrr turrrr*. **DISTRIBUTION** In Egypt fairly common summer visitor. Occurs in the Delta, Valley south to Abu Simbel, North Sinai and the Western Oases. Widespread on passage. Summer breeder over much of Europe and western Eurasia and North Africa. Winters in sub-Saharan Africa. Declining throughout its range due to changing agricultural practices and, especially in the Mediterranean, active hunting. **HABITAT AND HABITS** Open woodland, farmland, oases, and parks and gardens. Arrives late in spring and leaves early in autumn. Rather unobtrusive in tree canopy, but also seen perched on telegraph wires and feeding on the ground. **SIMILAR SPECIES** The Laughing Dove (Palm Dove) *S. senegalensis* is similar to the European Turtle Dove but smaller and slimmer, with a proportionally longer tail with more white – no black tip. It is more uniform above without the scalloping and with a blue-grey 'shoulder' panel and a gorget of fine black spots across the upper chest. Very common resident throughout Egypt's Delta, Valley, Western Oases, North and Red Sea coasts and Sinai. It is very familiar in villages, towns and cities throughout, even in central Cairo. The Eurasian Collared Dove *S. decaocto* is common in the Delta and Valley across to north Sinai, and appears to be expanding its range.

European Turtle Dove *Asha*; *Menut* is pigeon

This bird occurs in both sacred and secular contexts within Egyptian art. **PRACTICAL USES** Popular food until today; could possibly have been used in medicine and/or magic. **MYTHOLOGY/ICONOGRAPHY** Like all birds, the Turtle Dove was also a wild creature that needed to be controlled so that order (*maat*) was maintained. It, sometimes together with pigeons and other avifauna, was shown in poultry yards. It did play a small role in sacred contexts. The Coffin Texts contain a spell in which the deceased is transformed into a dove. At the same time, images of a flying dove hover over a pile of offerings in the wooden coffin of Djehutynakht from el-Bersha. The Book of the Dead also makes reference to Turtle Doves in Chapter 166 in connection with resurrection. Mummified food offerings of doves and pigeons are also known. **HIGHLIGHTS** The Memphite tombs contain images of doves being brought as offerings and being fattened in poultry yards (for example, in Saqqara, the tombs of Mereruka, Kaemnofret, Kagemni, and Ti, No. 60). Laughing Doves *S. senegalensis* also feature in many Memphite tombs, and Khnumhotep III's tomb (No. 3) at Beni Hasan contains images of Turtle Doves.

R. Pace

An offering of a Turtle Dove; Djehutynakht's coffin (MFA 20.1822)

Barn Owl ■ *Tyto alba* 33–39 cm WS 80–95cm

DESCRIPTION Medium-sized owl appearing very pale overall. Sandy-gold above with grey mottling (black and white speckling) on wings and crown. Underparts pure white sometimes tinged buff. White, heart-shaped facial disk. Eyes black. In flight appears all pale, ghostly white at night, with blunt wings – almost like a giant moth. **VOICE** Rattling shrieks and squeals. Young on nest utter an incessant and harsh hiss. **DISTRIBUTION** In Egypt a widespread resident including in the North Coast, Delta, Valley and Sinai – even in central Cairo. In region resident throughout and with a near global distribution. **HABITAT AND HABITS** Open country including palm and acacia groves, farmland, villages, towns and cities. Often associated with human activity, frequently nesting on buildings. Nocturnal. Often seen flying ghost-like at night, where can be confused with the Senegal Thick-knee *Burhinus senegalensis*, a nocturnal wader also found around human settlements, but note rounded, not pointed, wings. Pellets are more spherical than those of other owls. **SIMILAR SPECIES** Eight species of owl have been recorded in Egypt, with five as breeding residents, including the very rare Hume's Owl *Strix butleri*. The Pharaoh Eagle Owl *Bubo ascalaphus*, Egypt's largest owl at up to 50cm, has been recorded from much of the Eastern and Western Desert and Sinai. It can be identified by size, glaring orange eyes and prominent ear-tufts. Until recently it bred on the Step Pyramid at Saqqara and has been recorded at other temples such as Kalabsha south of Aswan. The diminutive Little Owl *Athene noctua* is also widespread – look for it at Abu Sir and Saqqara. At just 24cm, it can be identified by the rather square head, no ear-tufts (compare with scops owls *Otus* species) and glaring yellow eyes. It is also active by day. The Short-eared Owl *Asio flammeus* and Long-eared Owl *A. otus* are both migrants and winter visitors, with the latter perhaps now breeding in the region.

CC B. Brewer

Barn Owl 𓅓 *Em*

Barn owls appear mainly in hieroglyphic texts from as early as the First Dynasty as the letter 'm'. **PRACTICAL USES** None, although they may have been encouraged to control vermin. **MYTHOLOGY/ICONOGRAPHY** The letter 'm' is shown as an owl. **HIGHLIGHTS** Throughout Egypt the owl appears as the letter 'm'. In a scene of a papyrus swamp in the Theban tomb of Neferhotep (A.5) an owl is shown on a nest filled with eggs. Many sculptor's models or trial pieces from the Late Period and particularly the Ptolemaic Dynasty depict owls.

S. Ikram

Tomb of Khuwy, South Saqqara

Pied Kingfisher ■ *Ceryle rudis* L 25cm (including bill 5cm) WS 46cm

B. & T. O'Kane

DESCRIPTION Large, entirely black and white kingfisher. Adult black and white throughout; large headed with angular crest that is slightly shaggy at back. Underparts white with two black breast-bands in male, one in female. Tail quite long, largely white with black terminal band. Bill black and dagger-like. Feet and legs dark. In flight, upper wing chequered with outer wing black with white wing-patch. Flies fast and low over the water, hunting either from riverside branch or hovering high over the water before plunging down after prey. **VOICE** Vocal with loud, high-pitched wittering, especially when gathered in loose flocks. **DISTRIBUTION** In Egypt resident throughout the northern lakes, Delta and Valley as well as Western Oases. Elsewhere throughout the Middle East, sub-Saharan Africa and South Asia to China. **HABITAT AND HABITS** Tied to water. Along coasts, lakes, rivers and irrigation canals even in towns and cities, including central Cairo. Easily seen from cruise boats and feluccas. Hovering characteristic. Feeds almost entirely on fish. Breeds in burrow in sandbanks, sometimes colonially. **SIMILAR SPECIES** Two other kingfisher species are regularly recorded in Egypt. The White-throated Kingfisher *Halcyon smyrnensis* is a large, spectacular kingfisher, bright blue above with a chocolate-brown head and white throat and a spectacular red bill. It has undergone a dramatic range expansion in recent years in Egypt, and is now resident over much of the Delta and Valley south at least to Luxor. It is less tied to water than the Pied Kingfisher. The Common Kingfisher *Alcedo atthis* is smaller, greenish-blue above, with a bright cobalt back and reddish-orange underparts. Bill all black in male and black with red base in female. Winter visitor to fresh and saltwater habitats.

Pied Kingfisher *Anehbet*

Particularly common in images from the Old Kingdom. **PRACTICAL USES** None, or may possibly have been caught for sport and/or eaten. **MYTHOLOGY/ICONOGRAPHY** None known. **HIGHLIGHTS** Marsh scenes from King Userkaf's temple, now in the Egyptian Museum, depict a kingfisher. These birds feature frequently in Old Kingdom Memphite tombs (such as Hetepherakhti, D 60, at Saqqara), either in flight or on a nest (not an accurate depiction, as these birds nest in mudbanks and not in papyrus thickets). Painted scenes from the North Palace's 'green room' at Amarna (Egyptian Museum) show kingfishers hunting.

S. Ikram

Kingfishers protecting their nest; Tomb of Mereruka, Saqqara

Hoopoe ■ *Upupa epops* L 27cm (including bill 5cm) WS 46cm

DESCRIPTION Very distinctive black, white and cinnamon bird with a large crest. Upperparts pale orange-pink with mantle and wings boldly barred black and white. Belly white. Crest, which is orange-pink tipped black, normally held flattened back over crown but is erected on alighting or when alarmed. Bill long, slender and downcurved. Legs blue-grey. In flight appears very black and white with broad, rounded wings, white rump and black tail with broad white band. **VOICE** Song a rather deep sonorous *oop oop oop*. Alarm call *terrr* or *scheer*. **DISTRIBUTION** In Egypt, a common resident and widespread passage migrant throughout the North Coast, Delta, Valley, Western Oases and Sinai. In region, widespread resident and on passage. Ranges over much of southern Europe, South and East Asia, and Africa. **HABITAT AND HABITS** Farmland, orchards and groves, parks and gardens even in urban areas. Feeds on large insects such as grasshoppers and beetles, and other invertebrates. Feeds largely on the ground and may be seen foraging with a distinctive bobbing motion, probing the soil. Look out for Hoopoes in the grounds of hotels throughout the Delta and Valley, and also now in the gardens along the Red Sea and Sinai tourist villages. **SIMILAR SPECIES** None. The Hoopoe is sometimes divided into three species, the Eurasian Hoopoe described above, the African Hoopoe *U. africana* and the Madagascar Hoopoe *U. marginata*. These are sometimes placed in a family of their own, lumped in with the African woodhoopoes, or included with the hornbills and ground hornbills. A particular feature of the Hoopoe is its nesting habits. When brooding the female produces a foul-smelling secretion smelling of rotting meat, thought to deter predators and parasites, which she spreads over her plumage. This is produced by the nestlings, too, who further defend the nest against intruders by spraying faeces and hissing at the threat.

T. Robinson

Hoopoe *Djeb*

Hoopoes are frequently pictured in scenes depicting marshes and trees, as well as associated with children. **PRACTICAL USES** Hoopoes were kept as pets, as they still are today. They could be eaten as food and used in medicine and magic, notably to improve the eyesight and memory. In the Islamic tradition the hoopoe was the trusted bird of King Soliman, and allowed to carry his seal. **MYTHOLOGY/ICONOGRAPHY** Hoopoes sometimes feature in funerary papyri and on coffins in association with sycomore fig trees, sacred to the goddess Hathor in her guise as a deity associated with feeding the dead. Quite possibly the bird's crest evoked solar aspects of the deity, and the female Hoopoe's fierce protection of its nest echoed the goddess's protection of the deceased. **HIGHLIGHTS** A beautiful painted image of a Hoopoe in a tree exists in the tomb of Khnumhotep III (Tomb 3) at Beni Hasan, with several images of Hoopoes in the marshes or in association with trees found in the mastabas of Giza and Saqqara. These tombs also contain images of children, particularly boys, holding Hoopoes, for example the mastaba of Ptahhotep (D64) at Saqqara.

S. Ikram

A boy holding a pet hoopoe; Tomb of Mereruka, Saqqara

A. Gabr

Hoopoe on a papyrus umbel; Funerary Temple of Userkaf, Egyptian Museum

S. Ikram

Hoopoe on nest with eggs (unrealistic); Tomb of Inwmnw, Saqqara

Hooded Crow ■ *Corvus cornix* L 47cm WS 94cm

DESCRIPTION Medium-large, black and grey crow. Adult has grey mantle, nape and underparts. Head, wings and tail black with black extending down centre of breast. Bill moderately heavy. Legs and feet black. In flight, note grey underwing and undertail-coverts and rounded end to rather square tail. **VOICE** Vocal. Generally a harsh, abrasive *krrah*. **DISTRIBUTION** In Egypt a very common resident throughout the Delta and Valley, across the North Coast and Sinai. Recorded in the Western Oases. In region, in the Levant to Jordan but absent from Arabia; also much of Eastern Europe and Western and Northwestern Asia. **HABITAT AND HABITS** Widespread and common commensal found in most towns and cities, in villages and agricultural land, scrub and open country with trees. Sociable. Opportunistic feeder with a wide diet including scavenging on rubbish tips. Nest bulky and of sticks, normally high in a tree but also on cliffs and buildings. **SIMILAR SPECIES** Previously treated as conspecific with the Carrion Crow C. *corone* but now recognized as distinct. The Hooded Crow is the common crow of the Delta and Valley and other agricultural areas. It is replaced in desert and semi-desert habitats by the Brown-necked Raven C. *ruficollis*. This is similar in size but black throughout. The brown gloss on the neck is hard to make out except in good light. In flight, note the rather long, wedge-shaped rather than square tail. Often frequents rest houses on desert roads. The House Crow C. *splendens* is smaller at 43cm, with a longer bill and steep forehead. It is an invasive species probably introduced by ship, and is established and spreading along the Red Sea coast south from Suez. Can be very noisy and aggressive.

B. Grewal

Crow 𓎼𓅀𓃀𓎼𓅀𓄿𓅱 *Gabgaw*

Crow pecking out a fallen enemy's eyes; Battlefield Palette, British Museum (EA20791)

While crows must have been common in ancient Egypt, they are infrequently depicted except on ostraca and probably on papyri, which are now lost. **PRACTICAL USES** Regarded as a nuisance as they raided fruit trees, but used as a materia medica (for example to stop people from going grey). **MYTHOLOGY/ ICONOGRAPHY** Crows do not play a notable role in Egyptian mythology but were shown in scenes of daily life. The earliest image of crows shows them pecking at the dead on the battlefield (the Battlefield Palette, British Museum). Subsequent images are less gory, with crows stealing fruits, often from palm trees, or hanging about butchers' shops. There are two birds that have been tentatively identified as crows on a pectoral found at el-Riqqa, in place of the Horus Falcon that generally features on such objects. This might be due to an association of crows with a specific manifestation of Horus at the site of Leotopolis. **HIGHLIGHTS** Crows feature on collections of ostraca found in Egyptian and other museums, on the Turin 'Erotic and Satirical Papyrus'.

Red-backed Shrike ■ *Lanius collurio* L 18cm

DESCRIPTION Stout-billed, strongly marked passerine. Adult male chestnut-brown above with pale grey on nape, crown and forehead. Black 'bandit mask' from bill-base through eye to ear-coverts contrasting with white cheeks and throat. Underparts pale-washed pink. Pale grey rump, and tail black with white flashes. Female duller and browner, with brown mask and scaled pale underparts. Bill dark, short and stout. In flight, male with black tail with white bases to margins, chestnut mantle and no white on wings. Female also no white on wings, browner with white margins to dark tail. **VOICE** Song a scratchy warble, alarm call a repeated *tschek*. **DISTRIBUTION** In Egypt widespread passage migrant. Breeds over much of Europe south to the Levant. Widespread throughout the region on passage, and winters in sub-Saharan Africa south to South Africa. Generally declining. **HABITAT AND HABITS** On passage in farmland, olive groves, parks and gardens, orchards and scrub. Active predator, feeding largely on insects such as beetles and grasshoppers. Sometimes uses tree thorns and even barbed wire to impale surplus prey. **SIMILAR SPECIES** The larger Great Grey Shrike *L. excubitor* is a common resident in Egypt found in farmland, parks and gardens, groves and semi-desert country in the Delta, Valley, North Coast and Sinai. It is pale grey above with black wings and tail, and white underparts. In flight, white patch on wing and white tail margins. The Woodchat Shrike *L. senator* is common on passage and distinguished by the chestnut crown and nape. The Masked Shrike *L. nubicus* is also common on passage, especially in the east and Sinai. Male predominantly black and white, pale below flushed with orange, strongly so on the flanks. Female similar but overall paler.

B. Grewal

Red-Backed Shrike No known name in Egyptian

Rarely pictured in Egyptian art. **PRACTICAL USES** Conceivably eaten, otherwise none. **MYTHOLOGY/ICONOGRAPHY** None, save for being a bird that represents nature and wildness to be controlled. **HIGHLIGHTS** Only one example of this bird, perched in a tree, has been identified in Egyptian art, in the tomb of Khnumhotep III (No. 3) at Beni Hasan. The same tree contains an image of a Masked Shrike. Another example of the Masked Shrike is found in the tomb of Baqet III (No. 15) at Beni Hasan.

From Khnunmhotep III's tomb at Beni Hasan, watercolour by Howard Carter (Carter MSS.vii.1.5.2)

House Sparrow ■ *Passer domesticus* L 15cm

S. Fonseca

DESCRIPTION For many, the archetypal LBJ (Little Brown Job) – a small, streaky brown seed eater. Adult male brown, streaked dark above with white wing-bar. Underparts rather uniform pale grey. Throat and chest black, the extent of which varies. Crown grey with dark chestnut margins and pale grey cheeks. Bill short, stout and pointed. Legs pinkish. Female lacks the black on throat and breast, more uniform with pale buff supercilium. In flight, note white wing-bar. **VOICE** Varied chirps and cheeps. In the evening when gathering at regular roost sites in trees and bushes, large flocks can be very loud. **DISTRIBUTION** In Egypt a very common resident throughout, including in the centres of major towns and cities as well as remote oases, tomb and temple sites, and so on. Widespread resident in the region, except central Arabia. Common over much of Eurasia and widely introduced elsewhere. **HABITAT AND HABITS** Widely commensal. Found in all major towns and cities, in villages and agricultural land, and scrub and wasteland. Sociable, often occurring in flocks, roosting communally in trees even in city centres such as Cairo. Extent of black on chest may signify social status or merely age. **SIMILAR SPECIES** The Spanish Sparrow *P. hispaniolensis* is a widespread passage and winter visitor to Egypt to agricultural areas, farmland and gardens. The female is very similar to the female House Sparrow but often with obscure streaking below. The male can be distinguished by the chestnut-brown crown, narrow white supercilium, white cheeks and black breast extending to black streaking below and on flanks. It has recently been shown to breed in Siwa. The Tree Sparrow *P. montanus* is a very rare vagrant, smaller with chestnut crown and white cheeks with a bold black spot.

House Sparrow Kesenu

This very common bird was unpopular in ancient Egypt as sparrows attacked seeds and seedlings and thus were a major agricultural threat, and an enemy of the farmer. At its most positive, the sparrow appeared in the word small or narrow, but more commonly, it was used in negative ways, to denote bad, defective, empty, ill or diseased things. **PRACTICAL USES** None, although it may have been eaten in extremis as is the case in some countries even today. **MYTHOLOGY/ ICONOGRAPHY** None. Possibly regarded as bad or evil due to the use of its glyph in writing. **HIGHLIGHTS** Shown devouring sycomore figs, tomb of Userhet (No. 51), Thebes, as well as in texts.

K. Vertes

Tomb of Djehutymes (TT 295)

K. Vertes

Tomb of Nebamun (TT 179)

Egyptian Cobra ■ *Naja haje* L to 194cm in Egypt

R. N. Hewison

DESCRIPTION Long, relatively slender snake. Body rather uniform yellowish-brown, sometimes with yellow flecking. Below uniform pale yellowish with a broad, dark band across the neck. Head has broad, blunt snout and large eyes with round pupils. Neck can be flattened and flared into a 'hood' when threatened, when the snake raises the forepart of the body off the ground. Venom with toxin administered through fixed, hollow fangs, generally aiming at the eyes of a threat. Juvenile has black head and neck, and narrow, dark bars. **DISTRIBUTION** In Egypt recorded from the Northern Coastal Strip from the western Delta to Salum, and also from the Delta and Valley south to at least Edfu. Status around Lake Nasser uncertain. In North Africa, south Kenya, and west to Tunisia and Morocco; also occurs in south-west Arabia. The taxonomy is complex and the Arabian race is sometimes considered a distinct species. **HABITAT AND HABITS** Absent from true desert, but otherwise found in a variety of habitats, from semi-desert to agricultural areas, canal banks, ruins and old houses, and desert margins. In Egypt shows some variation, with animals from the Northern Coastal Strip generally darker than individuals from the Valley. Largely nocturnal or crepuscular, although it can be seen basking by day. Feeds on toads but also small mammals, birds, lizards and other snakes. **SIMILAR SPECIES** The Nubian Spitting Cobra *N. nubiae* is a species from the Nile Valley south from around Asyut to the shores of Lake Nasser and also Gebel Elba. As the name implies it can eject venom some distance at a threat. It is largely brownish-grey with two dark bands on neck and nape. A third cobra species, the Black Desert Cobra *Walterinnesia aegyptia*, is an uncommon, largely black species confined in Egypt to the Suez Canal Zone and parts of Sinai.

Cobra *Djet* or *Ikher*

Cobras are the most frequently depicted snake in Egyptian art throughout Egypt's long history. **PRACTICAL USES** Medicine, magic. **MYTHOLOGY/ICONOGRAPHY** Cobras are associated with a variety of deities and have both positive and negative connotations, depending on the context. They are linked to solar deities of creation, like Re and Atum, as they bask in the sun. The uraeus worn on the king's brow symbolizes his relationship with these creator gods, with the cobra providing extra protection to the king. Sometimes the cobra could be identified with the goddess Neith, protectress of the king as well as of Re. The cobra's abundance in the Delta makes it the manifestation of Wadjet, the titulary goddess of Lower Egypt. As such, the cobra, together with the vulture, is part of the royal titulary, as each animal is associated with the goddesses of Lower and Upper Egypt, respectively. They are thus featured on royal crowns from the later New Kingdom onwards. Because cobras are often found in fields, they are linked to Renenutet, the goddess of harvest and plenty. Cobras also represent underground, chthonic forces, and the power and magic of the deep earth. Like all snakes, cobras symbolize rebirth and regeneration as they shed their skins and are renewed and also because they hibernate and return from a state of quasi-death. Thus, they feature prominently in Egyptian funerary texts, with both good and bad cobras being present. Meretseger (she who loves silence), the goddess of the Theban mountains and thus of all the cemeteries there, is shown as a cobra emerging from the mountain. The goddess Weret-Hekau, Great of Magic, is also shown with a cobra head. **HIGHLIGHTS** Images of cobras abound in Egyptian art. A cobra adorns Tutankhamun's mask and his brow on all of his coffins. Most royal statues sport a uraeus. The funerary books depicted on the walls of the tombs in the Valley of the Kings contain cobras of different sorts.

S. Ikram

Tomb of Montuemhat, Deir el-Bahari

S. Ikram

The mask of King Tutankhamun with the uraeus and vulture on his brow, Saqqara

Horned Viper ■ *Cerastes cerastes* L Up to 73.5cm in Egypt

CC MinoZig

DESCRIPTION Stocky, medium to medium-large snake. Thickset with, like most vipers, a narrow neck and broad head. Scales along dorsum strongly keeled. These scales can be rubbed together when the snake is disturbed, producing a rasping, hissing sound. Generally pale beige above with a series of variable darker brown patches along back and smaller dark spots. Below, uniformly pale to white. Dark stripe from behind eye to gape. Eye has vertical pupil. 'Horns' are elongated supraorbital scales behind eye, and while distinctive are not always present. **DISTRIBUTION** In Egypt, including Sinai, widespread in a variety of desert habitats and absent only from the Delta and much of the southwestern Western Desert. Elsewhere in North Africa south to Sudan and across the Middle East. **HABITAT AND HABITS** A snake of the desert and semi-desert. It can be found in some of Egypt's most extreme deserts, but also in wadi floors, agricultural margins including oases, and mountainous, rocky country. Largely nocturnal. An ambush hunter, it shimmies down into the sand until entirely hidden save for the eyes and 'horns'. Preys on small mammals such as jirds, gerbils and jerboas, but in the Western Desert can survive on migrating birds during the spring and autumn migrations. In the desert, the alarm call of the White-crowned Wheatear *Oenanthe leucopyga* can indicate a Horned Viper. The snake is venomous with large, erectile hollow fangs and should be treated with caution. **SIMILAR SPECIES AND NOTES** The Sand Viper *Cerastes vipera* is similar in form to the Horned Viper but is smaller, lacks horns and although widely distributed in Egypt, is more confined to areas of sandy desert. Other vipers of the genera *Echis* and *Pseudocerastes* are found in Egypt.

Horned Viper *Fe-Kher-Debwy*

S. Ikram

Tomb of Montuemhat, Deir el-Bahari

The sign equivalent to the letter 'f' is the horned viper. These animals were very common in Egypt and feared for their venomous bite. **PRACTICAL USES** None known, although perhaps they had some medicinal or magical uses. **MYTHOLOGY/ ICONOGRAPHY** The symbol for the sound 'f'. The twelfth province (nome) of Upper Egypt, located on the east bank, north of the modern city of Asyut, was called the Mountain-of-the-Horned-Viper, either because part of the mountain looked like the viper, or because the area was densely populated by these snakes. **HIGHLIGHTS** The species can be seen in many texts in temples and tombs throughout the country. Depending on the state of preservation, the details are often carefully painted. It is also depicted as part of the province name in the White Shrine of Senusert I at Karnak.

J. M. Galán

Viper cut in two to render it harmless in the afterlife; Coffin of Iqer (TT11), Dhehuty Project

Nile Crocodile ■ *Crocodylus niloticus* L to 6m, but generally, and for all Egyptian records, smaller. WT to over 700kg, with female smaller

DESCRIPTION Huge reptile with long, slender snout, armoured body and long tail with serrated ridges. The archetypical crocodile. Heavy set and elongated. Snout elongated and with prominent teeth, visible at rest. Body heavy and armour plated. Tail long, laterally depressed and with serrated crest. Limbs powerful and clawed, with webbing especially on hindlimbs. Generally olive-brown above, banded dark, and uniform pale yellowish below. **DISTRIBUTION** Extinct in modern Egypt except for a population of up to 5,000 potential breeding adult crocodiles in Lake Nasser. Historically found throughout the Nile Valley and Delta and also in Fayoum. By the late nineteenth century largely confined to the Nile south of Aswan. The construction of the High Dam in the 1960s and '70s has created new habitat and the population in Lake Nasser is now established. Elsewhere throughout sub-Saharan Africa, including isolated populations in, for example, Sudan and Somalia. **HABITAT AND HABITS** Occasionally brackish or salt water but generally freshwater rivers and lakes, and in Egypt now restricted to Lake Nasser. Largely nocturnal but most often seen basking on sandbanks by day. Feeds on fish (such as Nile Perch), mammals and carrion. Lays eggs in an excavated nest burrowed in the sand. Displays much more elaborate parental care than most reptiles, with both sexes defending the nest against predators such as mongooses, foxes, jackals and monitor lizards. **SIMILAR SPECIES** No other crocodilians in Egypt. The Nile Monitor *Varanus niloticus* reaches a maximum length of c. 2m in Egypt. It lacks the armoured scaling and the tail's serrated dorsal crest of the Nile Crocodile and has a much shorter snout. Any 'crocodile' reported north of the Aswan Dam in modern Egypt is likely to be this species.

B. & T. O'Kane

Nile Crocodile *Meseh* or *Kapu* or *Depy*

S. Ikram

Nile Crocodile; Tomb of Ti, Saqqara

Crocodiles appear in Egyptian art from the Predynastic Period onwards and might have been first depicted in rock art. **PRACTICAL USES** Crocodiles could be eaten. Their hides could be used for armour and weapons, and their teeth in jewellery. Different parts of the crocodile were used in medicine and magic. Crocodile mummies were dedicated to the god Sobek. **MYTHOLOGY/ ICONOGRAPHY** Crocodiles played a complex role in Egyptian mythology. They could be manifestations of evil and a threat, and were to be killed and conquered. However, the god Sobek manifested as a crocodile or a man with a crocodile's head. Sobek was a god of fertility and sexual potency. He also had solar aspects – probably because crocodiles bask in the sun, their scutes or osteoderms look like sun bursts along a central ridge, the spikes on their tails evoke sun rays, and their eyes shine yellow at night, like mini suns. Sobek was revered particularly in the agriculturally rich area of Fayoum and in the Upper Egyptian area of Kom Ombo. Crocodiles were supposed to be able to predict the height of the Nile flood, crucial for agricultural wealth, as they knew where and at what level to make their nests and lay their eggs so that they would be safe from the inundation. **HIGHLIGHTS** A statue of Sobek and King Amenhotep III is to be found in the Luxor Museum. Mummified examples occur in the Animal Mummy Room of the Egyptian Museum, in the Grand Egyptian Museum and in the Graeco-Roman Museum in Alexandria. The Kom Ombo Crocodile Museum contains enormous mummified crocodiles as well as smaller, elaborately wrapped examples and crocodile eggs, and also statues, stelae and reliefs featuring crocodiles. Kom Ombo temple is dedicated to Sobek as well as to Horus.

S. Ikram

Nile Crocodile lying in wait to eat the Hippopotamus's baby and/or afterbirth; Tomb of Idut, Saqqara

Nile Mascarene Frog ■ *Ptychadena nilotica* (Formerly Mascarene Ridged Frog *P. mascareniensis*) L to 57–68mm

Shutterstock/Y. Oskanov

DESCRIPTION Medium-sized, slender frog with rather elongated and pointed snout. Upperparts greenish to brownish spotted dark, often with pale dorsal strip. Back and flanks have a series of six ridges. Underside whitish. Limbs have dark bands and yellow on back of thigh. Throat mottled. Long hindlimbs with extensive webbing between toes, which have thin, claw-like ends. Vocal. **DISTRIBUTION** In Egypt found throughout the Delta and much of the Valley. Also recorded from Fayoum and the Suez Canal zone. Status in Sinai unclear. Elsewhere, range extends through East Africa but status not clear in Sudan, South Sudan and Ethiopia. **HABITAT AND HABITS** The *Ptychadena* frogs are generally frogs of shallow, marshy habitats. In Egypt recorded from shallow waters such as rice fields, damp meadows, shallow irrigation canals and swamp margins. Active by day but also at night. Breeds in shallow water. **SIMILAR SPECIES** The Levant Green Frog *Rana bedriagae* is also widespread and common in Egypt, although it does not extend as far south. It is more aquatic than the Nile Mascarene Frog. The Sahara Frog *Pelophylax saharica* ranges across North Africa, but in Egypt is confined to Siwa oasis. In addition to the frogs, Egypt hosts four toad species distinguished by warty skin and prominent parotoid glands extending back from behind eye. The most widespread toad species in Egypt is the African Common Toad or Egyptian Toad *Sclerophrys regularis*, found throughout the Delta and Valley. It is abundant throughout the agricultural and farmland habitats, but also in parks and gardens even in towns and cities. It is active by night and its vocalizations are a familiar nocturnal accompaniment. The African Green Toad *Bufotes boulengeri* extends further along the North Coast and has been recorded from all the Western Desert oases.

Frog/Toad *Kerere* or *Pegeget*; tadpole is *Hefen*

The word for frog in Egyptian, as is the case with many other animals, is onomatopoeic. The word for tadpole is *hefen*. Frogs and tadpoles appear in hieroglyphic texts and frogs are shown in marsh scenes, particularly in the mastabas of Giza and Saqqara. **PRACTICAL USES** In extremis as food, in medicine and magic. **MYTHOLOGY/ICONOGRAPHY** Frogs are associated with Heket, the goddess of magic, as tadpoles seem to magically appear, then metamorphose into frogs. She is shown as a frog or as a woman with the head of a frog. Tadpoles are the symbol for 100,000 as they are so plentiful when hatched. In one of the Egyptian creation myths from Hermopolis (modern Ashmunein), the world was created by eight gods, four frog headed (male) and four snake headed (female). **HIGHLIGHTS** The mastabas of Mereruka, Kagemni, Niankhkhnum and Khnumhotep, Ti, Ptahhotep, Idut and many others contain depictions of frogs perched on reeds in marsh scenes. They also feature as amulets that can be seen in many museums. In temples in Thebes, tadpoles are shown at the bottoms of palm-ribs held by gods when offering the king millions of years of life and rule.

S. Ikram

Tomb of Kagemni, Saqqara

Upside-down Catfish ■ *Synodontis batensoda* L 50cm WT 1.5kg

CC Haps

DESCRIPTION At least four *Synodontis* species historically are recorded from Egypt with head shield, barbels and very prominent, serrated spines on dorsal and pectoral fins. The Upside-down Catfish is a rather stocky fish with armoured caudal shield (clearly seen in many pharaonic representations) and contiguous dorsal and adipose fins. Colour variable from greyish-blue to grey-green with dark underside; juvenile more boldly patterned (pictured). Barbels black. Tail-fin forked and with pointed tips. **VOICE** Also known as Squeaker due to sounds produced by contraction around the air bladder and by scraping the prominent spines together. **DISTRIBUTION** In modern Egypt status uncertain, but used to be found as far north as Cairo during inundation. More recently reported as unpopular in Lake Nasser due to spines damaging nets. Recorded over parts of the Blue and White Niles, and extending from Ethiopia through Chad to Senegal and Gambia. **HABITAT AND HABITS** Deeper fresh water but also inshore. Renowned for its habit of swimming upside down, hence the English name. The dark underside and paler dorsum and the position of the mouth reflects this. Feeds on algae, detritus, plankton and small invertebrates. **SIMILAR SPECIES** Other species recorded in Egypt popularly known as 'catfish' include the Mudfish *Clarias anguillaris*, which reaches up to 1m in length, lacks the armoured head and has a continuous dorsal through adipose fin, but without such prominent spines. The tail-fin is rounded rather than forked. It is a shallow-water fish that can bury itself in mud during droughts due to its ability to breathe atmospheric oxygen through specially adapted organs associated with the gills. *Heterobranchus longifilis*, or Sampa, is distinguished by its long barbels and discontinuous dorsal and adipose fins. It can reach 1.5m in length. The Electric Catfish *Malapterurus electricus*, also found in the Egyptian Nile, lacks a dorsal fin.

Catfish *Nar* or *Wehaw*

Catfish of many varieties were and remain common in the Nile, some growing to enormous sizes and being famed for their strength, to the extent that one of Egypt's earliest pharaohs was called 'Catfish'. **PRACTICAL USES** Food, medicine, and the bones could be used as tools. **MYTHOLOGY/ICONOGRAPHY** These are creatures that are part of the wild world, which need to be controlled and tamed. Their ability to survive in all kinds of conditions has led to their reputation of being very strong and resilient. As some types of catfish can survive in mud for long periods of time, then 'walk' to water sources, they are regarded as having magical powers. **HIGHLIGHTS** The Narmer Palette in the Egyptian Museum, which is not only a commemoration of the first time that Egypt was united under one king, but also the object that establishes the canon of Egyptian art and royal iconography, has a catfish (*Heterobranchus*) in the king's name. Different types of catfish (*Clarias*, and *Synodontis* and *Bagrus* species) are depicted in fishing or Nilotic scenes in the Memphite mastabas as well as tombs of the Middle and New Kingdoms throughout Egypt.

S. Ikram

The catfish indicative of king Narmer's name on the palette in the Egyptian Museum

S. Ikram

Synodontis *being caught on a hook; Tomb of Idut, Saqqara*

Nile Perch ■ *Lates niloticus* L to 2m WT to 70kg, with a suspect record from Lake Nasser of 175kg

DESCRIPTION Large to very large, freshwater fish, greenish to greyish-brown above, silvery below and with spinous fins and large mouth. The Nile Perch is one of the largest species of freshwater fish. Body robust and elongated, tapering to protruding jaw and rising to dorsal hump, with first set of dorsal fins with clear spines. Tail rounded. Eyes black with yellow outer ring. **DISTRIBUTION** In Egypt, historically found throughout the Nile Valley, including the Delta lakes, encompassing Maryout, and Fayoum. It seems that populations have been greatly reduced or are locally extinct. However, it is a commercial game fish in Lake Nasser in southern Egypt. The literature points to big declines since the construction of the High Dam. Elsewhere, the Nile Perch is found along many of the large rivers of Africa. **HABITAT AND HABITS** The Nile Perch is reliant on well-oxygenated fresh water, although Lake Maryout is brackish. The young fish are generally found in shallower water, while adults are found in cooler, deeper water. Adults feed mainly on fish, but also crustaceans and molluscs, while younger fish also take smaller species and plankton. **SIMILAR SPECIES** Many other freshwater fish species have been recorded from the Egyptian Nile. Three species of mullet *Mugil* species occur that are largely recorded from more brackish water. They are rather slim fish, often with longitudinal stripes. Tilapia, belonging to the family Cichlidae, are widespread freshwater fish along the Nile. They are deep-bodied fish with a full dorsal fin that extends across the back and with a full tail, and are an important commercial and food fish. Their taxonomy has changed but the Latin name is still in popular use.

Nile Perch *Inet*

Among the many fish shown in Egyptian art, Nile Perch are perhaps the most frequently depicted. **PRACTICAL USES** As food and as animal mummies. **MYTHOLOGY/ ICONOGRAPHY** The Nile Perch is often paired with tilapia. These fish are symbolic of rebirth and resurrection, and in marsh scenes are shown being harpooned by the tomb owner, symbolizing his triumph over death and his ability to live eternally. It is unclear why the perch has this role – perhaps due to its dark eyes with their bright yellow outer ring evoking the sun, and its ability to grow to a very large size? They are also associated with the goddess Neith. The significance of tilapia is more easily understood. These fish have a long dorsal fin that evokes the rays of the sun, and are mouth brooders, with the eggs or fry being carried in the mother's mouth, which she opens to emit the fry when they are ready – this seemed magical to the ancient Egyptians. **HIGHLIGHTS** Amulets of both fish are found in Egyptian jewellery and are scattered through museums in Egypt and elsewhere. Tombs from the Old Kingdom through the Late Period contain images of the tomb owner spearing Nile Perch and tilapia. These (and other) fish occur in fishing scenes, Nilotic scenes, and sometimes even on offering tables (Theban tomb of Menna, TT69).

S. Ikram

Nile Perch (front) and tilapia (back); Tomb of Menna (TT69), Luxor

S. Ikram

Menna harpooning tilapia and Nile Perch (TT69), Luxor

Migratory Locust ■ *Locusta migratoria* L 40–60mm

CC C. Kooyman

DESCRIPTION Locusts are grasshoppers of the Acrididae family famous for their populations periodically assuming plague proportions. The Migratory Locust is a large grasshopper with short, stout antennae. Head heavy with steep forehead. Occurs in solitary and gregarious phases. Adult solitary phase is brownish-olive with varying amounts of green with mottled forewings, and colour variable between sexes and with habitat vegetation. Hindlegs large and powerful, and often banded brown. Gregarious phase is mottled brownish and yellow. Nymphs of both phases lack fully developed wings and are more brightly coloured, with gregarious phase yellow to orange-spotted black. The Acrididae grasshoppers are distinguished from crickets and bush crickets by having the 'ears' or tympana on the side of the body rather than on the legs. **DISTRIBUTION** Widespread in Egypt in solitary phase. Powerful flier and can be found deep in the desert. In gregarious phase swarms are highly mobile. Feeds on vegetation, and gregarious phase can be a serious agricultural pest. Elsewhere with a virtually worldwide distribution in both the Old and New Worlds, including much of North Africa and the Middle East. **HABITAT AND HABITS**. Can be found in all habitats. Generally solitary but when populations build up, behaviour changes and swarms can erupt to plague proportions. **SIMILAR SPECIES** A number of other grasshopper species are found in Egypt. The Egyptian Grasshopper or Egyptian Locust *Anacridium aegyptium* is widespread in scrub and farmland. It is similar to the Migratory Locust but can be distinguished by blue and orange on the hindlegs and eyes with vertical black and white stripes. Does not form swarms. The Desert Locust *Schistocerca gregaria* may occur in Egypt in swarm years.

Locust *Senhem*

Tomb of Mereruka, Saqqara

S. Ikram

It is often difficult to differentiate between locusts and other grasshoppers, especially when the paint is missing. The latter are solitary and the former can be either solitary or gregarious. **PRACTICAL USES** Possibly as food. **MYTHOLOGY/ ICONOGRAPHY** Locusts are mentioned in funerary books such as the Pyramid and Coffin Texts and the Book of the Dead. In the Pyramid Texts the king flies to the sky as a locust, and in the Book of the Dead there is a palace called the field of locusts/grasshoppers. **HIGHLIGHTS** A locust/ grasshopper-shaped cosmetic container is located in the Grand Egyptian Museum, as well as other examples in museums in Egypt and worldwide. Images of locusts/grasshoppers perching on a plant are commonly found in marsh scenes in the Memphite tombs. A painted image is to be found in the tomb of Seti I (KV 17) in the Valley of the Kings on the left side of the descending corridor, at the bottom of a column of text.

Sacred Scarab ■ *Scarabaeus sacer* L 1.9–4.0cm

DESCRIPTION Large, robust and largely glossy black beetle. Popularly known as dung beetles. Carapace almost as broad as long with subtle ridges. Thorax broad. Head with a series of six projections. Robust legs with the front legs having a further four projections. Limbs and thorax fringed with hairs. These are thought to be adaptations to burrowing behaviour. Antennae not elbowed. Larva a large, whitish, 'C'-shaped grub. **DISTRIBUTION** Widespread and often abundant in Egypt, including at the major oases and Sinai. Elsewhere distributed over much of North Africa, southern Europe and the Middle East, south and east as far as India. **HABITAT AND HABITS** In Egypt, in agricultural and farmland habitats and coastal areas. Elsewhere, and especially in southern Europe, associated with sandy, coastal environments such as dunes and marshes. The Sacred Scarab is most renowned for its habit of rolling balls of dung. Adults roll balls of fine dung and deposit them underground. The female rolls a ball of dung and deposits a single egg within the ball, then buries the ball, after which the egg hatches and the larva feeds off the ball. This behaviour would have been very familiar to the ancient Egyptians. **SIMILAR SPECIES** The Sacred Scarab is a member of the Scarabaeidae family of beetles, which is one of the largest insect families, with an estimated 35,000 species currently recognized. A number of other scarab species occur in Egypt, but not all big black beetles are scarabs. Various species of Darkling Beetle (Tenebrionidae) have been recorded in Egypt from urban environments including Cairo, and from some of the most extreme deserts. These are from the family Tenebrionidae and can generally be distinguished by the longer and more slender legs and absence of projections on the head and forelimbs.

Scarab 𓆣 *Kheperer*

K. Amaury

Scarab in a sun disk in the Valley of the Kings

Scarabs are perhaps the quintessential creature associated with ancient Egypt, with their form lending itself to the most typical and popular souvenirs since as early as the seventh century BC, if not earlier. **PRACTICAL USES** Possibly in religious or magical rituals. Mummified as offerings to the sun god. **MYTHOLOGY/ICONOGRAPHY** Scarab beetles, using their hindlegs, roll herbivore dung into balls and lay an egg in them; they then abandon them. When an egg hatches, the larva eats the dung and emerges from it. The Egyptians equated the beetle rolling the ball with the scarab pushing the sun across the sky and over the horizon, and the fact that the baby beetles came from the ball (sun), further strengthened the solar association of creation, connecting the beetle to the creator god Atum, who begat children by himself. The dung beetle and its dung ball became symbols not only of Atum, but more so of the god Khepri, the manifestation of the sun god at dawn, and associated with rebirth and resurrection. **HIGHLIGHTS** Scarabs of all sizes, worn alone as amulets or incorporated into more complex jewellery designs, are found in museums throughout Egypt and the world. A giant scarab carved by command of Amenhotep III, a king who emphasized links to the sun god, is located near the Sacred Lake in Karnak Temple in Luxor.

S. Ikram

Giant scarab of King Amenhotep III, Karnak

K. Amaury

Khepri, scarab-headed god of the rising sun; Tomb of Nefertari, Luxor

Plain Tiger ■ *Danaus chrysippus* WS 7–8cm (African Monarch)

DESCRIPTION Large, striking black, white and orange butterfly. Sexes largely similar. Body black with white spots. Upperwing mainly light orange-brown with black margins and darker tips to forewings, with bold white patches forming a band. Three black spots on otherwise uniform hindwing. Underwing with yellower hindwings with black spots. In male only, with white centre to main dark spot, which is a scent gland. **DISTRIBUTION** Widespread and often abundant in Egypt, including in the major oases and Sinai. Both resident and migrant. Elsewhere distributed over most of Africa and southern Europe, across the Middle East, to tropical Asia, reaching Australia. **HABITAT AND HABITS** Semi-desert and desert margins, scrub and bushland, agricultural land, parks and gardens. Main flight period in Egypt June–September. Strong flier active by day. Adult feeds on nectar. Caterpillar mainly pale grey boldly marked with black and yellow stripes and with six black 'tentacles', and reaches 40mm in length by the fifth instar. Feeds on toxic plants, in Egypt especially *Calotropis* species (milkweeds). The accumulation of toxins renders the butterfly unpalatable and the bold patterning and colouration act as a warning to predators. **SIMILAR SPECIES** The Diadem *Hypolimnas misippus* is a much rarer species in Egypt. The male is very distinctive, being all black with a large white patch on each forewing and hindwing margined with blue. Female is almost identical to the Plain Tiger but has only one black spot on the upperside of the hindwing as opposed to three. It gains protection by mimicking a toxic species. The Painted Lady *Vanessa cardui* is also dark and orange but much more muted and with more complex patterning. It is highly migratory and can be seen virtually anywhere in Egypt, including in remote desert such as the Gilf Kibir.

CC Haneesh K M

Plain Tiger No Egyptian name known

This is the only butterfly that is represented in Egyptian art. **PRACTICAL USES** None known. **MYTHOLOGY/ ICONOGRAPHY** The main association is that of rebirth and resurrection, as the caterpillar metamorphoses into a butterfly, with the pupa/cocoon stage paralleling the mummy. In addition, the colour and patterning of the wings evokes solar iconography. As the main foods of this butterfly are Giant Milkweed *Calotropis procera* and Crown Flower C. *gigantea*, both used particularly in medicines related to heart disease, it is possible that the butterfly was associated with strong hearts and enduring life. In addition, the Crown Flower was used in perfumes dedicated to the goddess Hathor, and thus, by extension, the butterfly might be associated with that goddess. **HIGHLIGHTS** Unpainted as well as painted images of the Plain Tiger are common in the marsh scenes in mastabas and in Eighteenth Dynasty Theban tombs, with a few examples found in Beni Hasan and Meir. Butterflies also appear on the bracelets of Queen Hetepheres, now in the Grand Egyptian Museum.

S. Ikram

Butterfly from the tomb of Kagemni, Saqqara

S. Ikram

Butterfly with folded wings; Tomb of Nebamun, British Museum (EA 37980)

S. Ikram

Painted butterfly; Tomb of Nebamun, British Museum (EA 37980)

Scorpions ■ Order Scorpiones Length from 1.3cm in *Microbothus pusillus* to more than 20cm in the tropical genus *Heterometrus*

DESCRIPTION A scorpion's body is divided into two main sections, the prosoma, covered by a single carapace, and the abdomen. The latter has a base of seven broad segments, then a flexible 'tail' that ends in a 'sting'. Scorpions have four pairs of legs, plus a pair of leg-like pedipalps armed with 'pincers'. Scorpions are renowned for their sting. Indeed, one Egyptian species, the African Golden Scorpion *Leiurus quinquestriatus* (pictured) is recorded as the most venomous of all scorpions, but the amount of venom injected is very small. **DISTRIBUTION** Scorpions are popularly associated with deserts and in Egypt they have been recorded from all desert regions, including extreme parts of the Western Desert. They are also found in oases, acacia groves, agricultural areas, and camps and villages, and are common on the shores of Lake Nasser. **HABITAT AND HABITS** Many scorpion species are highly adapted to the desert. All are nocturnal and spend the day in burrows under rocks and boulders. Many are active predators, with the African Golden Scorpion recorded climbing trees and bushes in search of prey, which is generally arthropods. **SPECIES** A number of scorpion species occur in Egypt. The Large-clawed Scorpion *Scorpio maurus* is an amber species with distinctly large claws. The Egyptian Sand Scorpion *Buthacus leptochelys* is similar but paler, with much more slender claws and tail. The Fat-tailed Scorpions *Androctonus australis* and *A. crassicauda* are darker than the two previous species and range across North Africa and the Middle East. The tail is much thicker. Also dark to black is the Egyptian Pillar-tailed Scorpion *Orthocirus innesi*, some 3–4cm long, with small, weak claws. It is distinctive in that the tail is held almost bent double over the back rather than vertically.

CC Minozig

Scorpion Djaret or Wehat

Scorpions are rarely depicted in tomb or temple art, save in conjunction with the goddess Selket/Serket (see below). They were regarded as powerful creatures, and one of Egypt's earliest kings was called Scorpion. **PRACTICAL USES** Possibly used in medico-magical contexts. **MYTHOLOGY/ICONOGRAPHY** Associated with Selket or Serket, the scorpion goddess or *Srket-htw*, She who Relieves the Windpipe, referring to undoing the paralysis and lack of breath that scorpion stings can engender. The goddess was capable of healing venomous bites and stings of scorpions, snakes and possibly minor insects. She is shown as a woman with a scorpion on her head, although in some early instances (c. 3050 BC) she is shown as a scorpion with a woman's head. She protected the dead and was associated with one of the sons of Horus, Qebehsenuef, who, together with Selket, protected the canopic jar containing the deceased's intestines. Another deity who protected against scorpion, snake and wild animal bites/attacks was the young Horus, who is shown facing frontally on stelae called *cippi* (*cippus*, singular), standing on crocodiles, and holding a variety of venomous creatures, including scorpions, in his hands. The stelae are densely inscribed with texts that protect against stings and bites. Priests would chant the spells, pour a liquid over the image and texts, thus accruing the magic, then people would drink the liquid, either as a cure for a bite or sting, or as a preventative measure. They could also anoint their bodies with the liquid. **HIGHLIGHTS** Scorpion amulets can be found in many museums. Images of Serket abound, with some fine examples in Edfu Temple, Nefertari's tomb, and as one of the guardian statues protecting Tutankhamun's canopic chest. Scorpions feature on King Scorpion's mace head, which is in the Ashmolean Museum in Oxford.

S. Ikram

Cippus *showing the child Horus standing on crocodiles and holding scorpions and other venomous creatures*

S. Ikram

A scorpion on the head of the goddess Selqet; Tomb of Nefertari

Below is a list of some of the key sources used in researching this book and other sources, for those looking to discover more about Egypt's wildlife, ancient and modern.

For those with a particular interest in birds, a valuable source of information on the birds of Egypt and across the Middle East is the Ornithological Society of the Middle East, the Caucasus and Central Asia (OSME). OSME's journal is *Sandgrouse*. It is published twice a year and is a great source for those wanting more specialist research. Its website is www.osme.org.

Aly, Dina & Khalil, Rafik. 2011. *Wildlife in South Sinai*. Dina Aly and Rafik Khalil, Cairo.

Aulagnier S. et al. 2009. *Mammals of Europe, North Africa and the Middle East*. A & C Black Publishers, 2009.

Baha El Din, Sherif. 2006. *A Guide to the Reptiles and Amphibians of Egypt*. AUC Press, Cairo.

Beaman, Mark & Madge, Steve. 1998. *The Handbook of Bird Identification for Europe and the Western Palearctic*. Christopher Helm, A & C Black, London.

Brewer, Douglas, J. & Friedman, Renée, F. 1990. *Fish and Fishing in Ancient Egypt*. AUC Press.

Bunbury, Judith, Cooper, John, P., Hoath, Richard, Ikram, Salima, Johnston, Christine & Schneider, Thomas. 2023. 'The Egyptian Nile: Human Transformation of an Ancient River'. In: Wantzen, K. M. (ed.): *River Culture – Life as a Dance to the Rhythm of the Waters*. pp. 43–77. Paris: UNESCO Publishing, Paris. DOI: 10.54677/MDJN3102. https://unesdoc.unesco.org/ark:/48223/pf0000382775?posInSet=1&queryId=f565d8d4-e146-4f67-8fa0-64803b3ebac7

Castelló, José R. 2016. *Bovids of the World*. Princeton University Press, Princeton.

Castelló, José R. 2018. *Canids of the World*. Princeton University Press, Princeton.

Clutton-Brock. 1988. *The British Museum Book of Cats: Ancient and Modern*. British Museum Publications, London.

Germond, Philippe & Livet, Jacques. 2001. *An Egyptian Bestiary: Animals in Life and Religion in the Land of the Pharaohs*. Thames & Hudson, New York.

Gilbert, Francis & Zalat, Samy. 2007. *Butterflies of Egypt: Atlas, Red Data Listing and Conservation*. Egyptian Environmental Affairs Agency, Cairo

Goodman, Steven M. & Meininger, Peter L. (eds). 1989. *The Birds of Egypt*. Oxford University Press, Oxford.

Hoath, Richard. 2009. *A Field Guide to the Mammals of Egypt* (2nd edition). AUC Press, Cairo.

Hoath, Richard. 2021. *The Birds of Egypt and the Middle East*. AUC Press, Cairo.

Hollom, P. A. D. et al. 1988. *Birds of the Middle East and North Africa*. T. & A. D. Poyser, Calton.

Houlihan, Patrick F. 1996. *The Animal World of the Pharaohs*. AUC Press, Cairo.

Houlihan, Patrick F. & Goodman, S. M. 1988. *The Birds of Ancient Egypt*. AUC Press, Cairo.

Ikram, S. 2020. 'Ancient Egyptian Fauna', *Oxford Handbook of Egyptology*, Shaw, I. &

Bloxham, E. (eds). Oxford: Oxford University Press. 151–64.

Ikram, Salima. 2015. *Divine Creatures: Animal Mummies in Ancient Egypt.* AUC Press, Cairo.

Ikram, Salima & Nasry Iskander. 2002. *Catalogue Général of the Egyptian Museum: Non-human Mummies.* Cairo: Supreme Council of Antiquities Press.

Ikram, Salima, Aicardi, Sara & Faccetti, Federica (eds.). 2024. *The Animal Mummies of the Museo Egizio, Turin.* Studi del Museo Egizio 5. Turin: Museo Egizio.

Jonsson, L. 1992. *Birds of Europe with North Africa and the Middle East.* Christopher Helm, London.

Lieske, Ewald & Myers, Robert F. 2004. *Coral Reef Guide: Red Sea.* HarperCollins, London.

Osborn, Dale J, & Helmy, Ibrahim. 1980. 'The contemporary land mammals of Egypt (Including Sinai)', *Fieldiana Zoology*, Field Museum of Natural History, New Series, No. 5. Chicago: Field Museum of Natural History.

Osborn, Dale J. & Osbornova, Jana. 1998. *The Mammals of Ancient Egypt.* Warminster: Aris and Philips.

Ottoni, C., Van Neer, W., De Cupere, B., Daligault, J., Guimaraes, S., Peters, J., Spassova, N., Prendergast, M. E., Boivin, N., Morales-Muniz, A., Balasescu, A., Becker, C., Benecke, N., Boroneant, A., Buitenhuis, H., Chahoud, J., Crowther, A., Llorente, L., Manaseryan, N., Monchot, H., Onar, V., Osypinska, M., Putelat, O., Morales, E. M. Q., Studer, J., Wierer, U., Decorte, R., Grange, T., & Geigl, E.-M. 2017. The Palaeogenetics of Cat Dispersal in the Ancient World. *Nature Ecology and Evolution* 1, 0139. Doi:10.1038/s41559-017-0139.

Parkinson, Richard. 2008. *The Painted Tomb-Chapel of Nebamun.* AUC Press, Cairo.

Porter, Richard & Aspinall, Simon. 2010. *Birds of the Middle East* (2nd edition). Helm Field Guides, Christopher Helm, London.

Redman, Nigel, Stevenson, Terry & Fanshawe, John. (2009) *Birds of the Horn of Africa; Ethiopia, Eritrea, Djibouti, Somalia and Socotra.* Princeton University Press, Princeton.

Sterry, Paul. 2000. *Collins Complete Guide to Mediterranean Wildlife.* HarperCollins, London.

Svensson, L., Mullarney, K. & Zetterstrom, D. 2009. *Collins Bird Guide* (2nd edition). HarperCollins, London.

Index of Common Names

Addax 78

Baboon, Hamadryas 38
Barn Owl 124

Catfish, Upside-down 144
Cheetah 64
Cobra, Egyptian 136
Cormorant, Great 100
Crane, Common 118
Crocodile, Nile 140
Crow, Hooded 130

Domestic Cat 58
Domestic Dog 46
Dromedary 72

Egret, Little 104

Flamingo, Greater 108
Fox, Red 42
Frog, Nile Mascarene 142
Fruit Bat, Egyptian 36

Gazelle, Dorcas 80
geese, true 92
Giraffe 74
Goose, Egyptian 94

Hare, Cape 84
Hedgehog, Long-eared 32
Hippopotamus 70
Hoopoe 128
Hyena, Striped 54

Ibex, Nubian 82
Ibis, Sacred 106

Kestrel, Common 114
Kingfisher, Pied 126

Lapwing, Northern 120
Leopard 60
Lion 62
Locust, Migratory 148

Mongoose, Egyptian 52
Monkey, Grivet 40
Mouse, House 86

Night Heron, Black-crowned 102

Oryx, Scimitar-horned 76
Ostrich, Common 88
Otter, Eurasian 50

Perch, Nile 146
Pintail, Northern 96

Quail 98

Scarab, Sacred 150
scorpions 154
Shrew, African Giant 34
Shrike, Red-backed 132
Slit-faced Bat, Egyptian 36
Sparrow, House 134
Swamphen, Purple 116
Swan, Mute 90

Tiger, Plain 152
Turtle Dove, European 122

Viper, Horned 138
Vulture, Egyptian 110
  Lappet-faced 112

Wild Ass, African 66
Wild Boar 68
Wild Cat, African 56
Wild Dog, African 48
Wolf, African Golden 44

Index of Scientific Names

Acinonyx jubatus 64
Addax nasomaculatus 78
Alopochen aegyptiaca 94
Anas acuta 96
Anser species 92

Camelus dromedarius 72
Canis lupaster 44
Canis lupus familiaris 46
Capra nubiana 82
Cerastes cerastes 138
Ceryle rudis 126
Chlorocebus aethiops 40
Corvus cornix 130
Coturnix coturnix 98
Crocidura olivieri 34
Crocodylus niloticus 140
Cygnus olor 90
Danaus chrysippus 152

Egretta garzetta 104
Equus africanus 66

Falco tinnunculus 114
Felis catus 58
  lybica 56

Gazella dorcas 80
Giraffa camelopardalis 74
Grus grus 118

Hemiechinus auritus 32
Herpestes ichneumon 52
Hippopotamus amphibius 70
Hyaena hyaena 54

Lanius collurio 132
Lates niloticus 146
Lepus capensis 84
Locusta migratoria 148
Lutra lutra 50
Lycaon pictus 48

Mus musculus 86

Naja haje 136
Neophron perenopterus 110
Nycteris thebaica 36
Nycticorax nycticorax 102

Oryx dammah 76

Panthera leo 62
  pardus 60

Papio hamadryas 38
Passer domesticus 134
Phalacrocorax carbo 100
Phoenicopterus roseus 108
Porphyrio porphyrio 116
Ptychadena nilotica 142

Rousettus egyptiacus 36

Scarabaeus sacer 150
Scorpiones order 154
Streptopelia turtur 122
Struthio camelus 88
Sus scrofa 68
Synodontis batensoda 144

Threskiornis aethiopicus 106
Torgos tracheliotus 112
Tyto alba 124

Upupa epops 128

Vanellus vanellus 120
Vulpes vulpes 42

▪ Acknowledgements ▪

Any guide of this type builds on the works that have gone before it; we acknowledge those works in the References and Further Reading pages and have had the privilege of having met and benefited from a number of those who have authored these works. Special thanks go to Neil Hewison. At John Beaufoy Publishing we would like to thank Rosemary Wilkinson and Krystyna Mayer for all their work and assistance, and at AUC Press Nadia Naqib for her advice and additional editing. For their special photos we want to specifically acknowledge Andy Main, Bernard O'Kane, Tracy Robinson, Ana Oliveira and José Galán. Devon Murray has been superb in her support and help with images, and Bianca van Sittert has done beyond a sterling job with all the hieroglyphs for the animal names. We are also indebted to the Griffith Institute for its generosity and help in providing the image from Khnumhotep III's tomb.

RH: I have had the joy and privilege of working with my co-author Salima. Both descriptives at the same time fit, and very much understate.

Thank you too to Yasmine Motawy, with whom I have worked on other projects, and who gave invaluable advice and wise counsel particularly at the beginning of this project.

For their support over the time working on this book I would especially like to thank Liz Hoath, Gretchen McCullough, Kathleen Saville, Dina Rateb, Melanie Carter, Peter Barsoum and Neil Hewison.

Finally, I would like to very gratefully acknowledge the late Pam Hoath for opening her home to an author's 'hub', something which she had done on numerous occasions before, but never more special than for this.

SI: As ever, it has been a delight to work with Richard, who did a lot of the heavy lifting. I hope we have a chance to do another project soon.